Agricultural and Forestry Reconstruction After the Great East Japan Earthquake

Toshiyuki Monma • Itsuo Goto
Takahisa Hayashi • Hidekiyo Tachiya
Kanju Ohsawa

Editors

Agricultural and Forestry Reconstruction After the Great East Japan Earthquake

Tsunami, Radioactive, and Reputational Damages

Editors

Toshiyuki Monma
Department of International
  Biobusiness Studies
Tokyo University of Agriculture
Tokyo, Japan

Takahisa Hayashi
Department of Bioscience
Tokyo University of Agriculture
Tokyo, Japan

Kanju Ohsawa
Tokyo University of Agriculture
Tokyo, Japan

Itsuo Goto
Department of Applied Biology
  and Chemistry
Tokyo University of Agriculture
Tokyo, Japan

Hidekiyo Tachiya
Soma City Office
Fukushima, Japan

ISBN 978-4-431-55557-5   ISBN 978-4-431-55558-2   (eBook)
DOI 10.1007/978-4-431-55558-2

Library of Congress Control Number: 2015947340

Springer Tokyo Heidelberg New York Dordrecht London

Printed on acid-free paper

Springer Japan KK is part of Springer Science+Business Media (www.springer.com)

Tokyo University of Agriculture's Philosophy and the East Japan Assistance Project

Introduction

Two and a half years have now passed since the Great East Japan Earthquake, but a range of challenges still hamper the revival of the agriculture, forestry, and fishery industries in the area affected, and circumstances remain difficult. The revival of these industries is particularly delayed in Fukushima Prefecture, where tsunami damage and radioactive contamination resulting from the Fukushima nuclear accident were compounded by reputational damage as a result of the contamination.

On behalf of Tokyo University of Agriculture, I visited the city of Soma in Fukushima Prefecture immediately after the earthquake and ensuing disasters, and promised the mayor, Hidekiyo Tachiya, that the university would join forces with Soma to restore the city's devastated agricultural and forestry industries. Since then, the university as a whole has implemented a wide range of initiatives to help with the revival of Soma's community. Here at Tokyo University of Agriculture we consider it our duty to nurture talented individuals who can use practical science to contribute to the development of agriculture, forestry, and fisheries. It seemed only natural, therefore, that we should get involved in efforts to restore the disaster-stricken region.

Tokyo University of Agriculture was founded 122 years ago by Takeaki Enomoto, an influential figure in the government that ruled Japan after the Meiji Restoration in 1868. As Japan's largest agricultural university, the university has dedicated itself to developing skilled young people, producing a steady flow of high-caliber individuals for employment across the farming industry. The university's educational philosophy is rooted in the practical-science-based ethos espoused by its first president, Tokiyoshi Yokoi. Convinced that scientific principles are meaningful only when put into practice, he advocated the need to "Ask the rice plant about the rice plant; ask farmers about agriculture." Thus, Yokoi always maintained that the key to researching agricultural science was to understand that theory exists only through practical application, and that practical application occurs only in conjunction with theory.

The university's educational policy is above all founded on this all-pervasive ethos of practical science. Moreover, it is founded on a pragmatic education that equips our students to adopt roles of responsibility in agriculture and forestry, enabling them to contribute to the development of communities as local leaders both in Japan and overseas. We therefore regarded it as our calling to use the university's accumulated knowledge and technological know-how to help with recovery following these major disasters.

Outline of the Project

In light of our university's founding purpose and philosophy, we believed strongly that we should do all we could to contribute to rehabilitating agriculture and forestry in the wake of the Great East Japan Earthquake. We decided to act without delay, enlisting the help of all teaching staff to confirm that our students were safe immediately after the disasters. These efforts revealed that about 1,500 of our students came from the areas affected, and the family homes of 250 of these students had sustained major damage, including total or partial destruction. At the beginning of April, having decided on measures to assist the students whose homes had been damaged and obtained the approval of the university's board of trustees, we launched the Tokyo University of Agriculture East Japan Assistance Project with the aim of rehabilitating the agriculture devastated by the disasters.

First, we appointed a project leader: Toshiyuki Monma, an expert on agriculture in the Tohoku region, which was worst hit by the March 2011 disasters. On behalf of the university, he investigated from a range of perspectives what could be done where. Based on his investigations, we decided to start doing what we could in the city of Soma in Fukushima Prefecture as part of a project to assist the region. During the series of public holidays starting at the end of April, I visited the area with 14 teaching staff and researchers. When we arrived, local guides showed us fields covered in debris and sludge, and the sight left us speechless. In the face of such destruction, what we as a university needed above all was assistance from the local community, and we decided that it would be essential to conduct our activities in liaison with Soma's local authority, the Fukushima Agricultural Technology Centre and its Hama District Research Institute, and the local agricultural and forestry cooperatives.

With the assistance of local residents, the project's research teams started their investigations, each focusing on a separate area, such as farming, farmland reconstruction, soil and crops, forest reconstruction, community revitalization, or reputation-based damage. It soon became apparent that the recovery faced a number of challenges. Above all, farmers affected by the disaster felt differently about whether to resume agriculture depending on how much damage their farms had sustained and the size of their operations. Many small-scale farmers had lost much of their motivation to farm, while the desire to resume operations had also declined among large-scale farmers whose farms had sustained major damage. Given such circumstances, it seemed a major challenge to rehabilitate farmland covered in debris, sea salt, and sand (although we were later able to demonstrate that application of slag from steelmaking converter furnaces effectively improved soil quality in paddy fields). We also suspected that radiocesium absorbed by trees would result in residual radiocesium in the forests for many years to come, because radiocesium absorbed by trees gradually works its way further inside the trees: that is an issue that continues to be of concern. In addition, we are still some way from fully understanding other issues that present a greater challenge still, such as economic damage from the negative reputation of Fukushima's agricultural and food products. Many radioactive "hot spots" have emerged from the effects of the cesium, and as a result farmers are beset by doubts about whether any agricultural products they produce will actually sell. What is needed, therefore, is an effective monitoring system to make absolutely certain no contaminated agricultural products are sold. We also need to decontaminate and restore the farmland that sustained major damage from the tsunami and identify effective methods of decontaminating the forests.

It is clear that it will take a very long time to rehabilitate the farmland and forests and to revive the community. On the most fundamental level, agricultural science is about creating the right environment for living organisms and cultivating a wide variety of plants and animals. In this case, we as a university still have a long battle ahead of us to find solutions to issues that have never before been experienced.

I sincerely hope that the knowledge and scientific technologies the university has cultivated since its founding will prove to be of use in restoring the agricultural and forestry industries in the disaster-stricken city of Soma in Fukushima Prefecture.

Chairman of the board of directors   Kanju Ohsawa, Ph.D.
Tokyo University of Agriculture
Tokyo, Japan

Foreword I

The earthquake off the Pacific coast of Japan's northeastern Tohoku region struck at 2:46 P.M. on Friday, March 11, 2011. It registered magnitude 9.0 with a maximum seismic intensity of 7 on the Japanese scale, and in Tokyo the seismic intensity reached upper 5. Here at the university, I was in a faculty meeting where we were just about to hear speeches from retiring teaching staff. We felt the ground shaking, and assumed that it would soon die down as it usually did, so nobody moved from their seats. But instead the shaking grew stronger, the windows began to rattle ominously, and as I wondered whether the display boards suspended from the ceiling might fall, the worst case scenario crossed my mind and a feeling of dread coursed through me.

When the shaking died down we were naturally very relieved, and all the teaching staff left the building, returning to their seminar rooms to lead the evacuation of their students to the sports ground. There were more than 1,000 students in seminar rooms or participating in club activities that day; teaching staff checked that they were all unharmed and led them to the sports arena to escape the cold. Then we were kept busy with a whole succession of demands such as distributing emergency food supplies and finding accommodation for students, teaching staff, and administrative staff who could not return home, as well as checking on the well-being of students who were visiting their families in the areas most affected by the earthquake.

Yet we never imagined that one hour after the earthquake a huge tsunami would strike the Pacific coast including the prefectures of Iwate, Miyagi, and Fukushima. We were lost for words when we saw the extent of the damage relayed to us on television, and thought of the sorrow and helplessness the people in the stricken areas must be experiencing. We were filled with a profound sense of loss at the formidable power of the tsunami as it stole so many precious lives, swept away houses, and deprived families of each other's company, destroying the region's farmland along with the very infrastructure of daily life. And there was more to come: the tsunami had caused a failure in the power supply at Fukushima Daiichi Nuclear Power Station, making it impossible to continue cooling the reactors. The resulting hydrogen buildup triggered explosions that caused partial collapse of the reactor buildings, releasing radionuclides that scattered and leaked across the surrounding areas where people were going about their daily lives. Thus, this area that had already been devastated by the earthquake and tsunami suffered still more tragedy.

On behalf of the university I wish to express my condolences for every life that was so abruptly taken, and extend my deepest sympathies to all who lost family members or were otherwise affected by these disasters.

Here at the university we continued to check on the well-being of our students and their families. To ensure everybody's safety amidst the ongoing aftershocks, we canceled the graduation ceremony at which students are awarded their degrees, as well as the university entrance ceremony, both of which are usually very well attended by both students and their families.

As time went by, we wondered how Tokyo University of Agriculture could take advantage of its position as an established center for agricultural science to help recovery in the afflicted areas. The unimaginable scale of the devastation only aggravated the frustration we felt at having to sit by and watch. In May, my predecessor as president, Kanju Ohsawa, traveled to the region affected by the disasters, and after observing the situation there in great detail, went on to set up the Tokyo University of Agriculture East Japan Assistance Project. It was now that the university would be called to act upon the words of its first president, Tokiyoshi Yokoi, who warned students against excessive reliance on a theoretical approach to agriculture, cautioning that "where agricultural science flourishes, farming itself fails."

Hidekiyo Tachiya, the city of Soma's mayor, provided his full support for the project, and in 2012, the year after the disasters, we embarked on the first step in the process of restoring the paddy fields damaged by the tsunami. Prompted by local farmers' renewed enthusiasm for the idea of reviving the area's agriculture, we planted rice in 1.7 ha of paddy fields using the Soma (or Tokyo University of Agriculture) Method, and approximately 10 tons of rice was harvested. Promoted as "Soma Revival Rice," it became a beacon of hope for agricultural reconstruction in the Soma area. In 2013 we went on to plant an area of 50 ha, and 200 tons of rice was harvested; the plan is to extend the area to 200 ha in 2014.

In addition to the Soma Revival Rice project, many other initiatives to revitalize agriculture and the local community are now gradually producing results. This book is a record of the efforts by the city of Soma and Tokyo University of Agriculture to bring about such revitalization in the disaster zone during the past two and a half years. I hope that the publication of this book will be a catalyst for the reconstruction of agriculture in the Soma area, and that it will also be helpful for reviving agriculture in other areas.

I wish to express my heartfelt gratitude to Soma's mayor and all the city's residents for their understanding and support during the progress of this project. Finally, let us raise a cheer for the passionate commitment to reviving Soma's agriculture among the university's teaching staff and students as they continue to play an active part in the project.

President, Tokyo University of Agriculture   Katsumi Takano, Ph.D.
Tokyo, Japan

Foreword II

The Great East Japan Earthquake hit on March 11, 2011, and left Soma's farmland in an appalling state.

The salt and sludge carried inland by the tsunami were accompanied by the wreckage of homes reduced to debris, huge fallen pine trees, and other waterborne objects that wreaked devastation as far as the eye could see across the once-beautiful paddy fields. The plastic greenhouses used for horticulture—particularly those used to grow strawberries—were all swept away and filled with sludge. The tractors, or in many cases, the homes, of Soma's farmers were washed away, and in the midst of such tragic circumstances, just when they were wondering how they could go on, the nuclear disaster delivered the final blow.

Then there was not only the tsunami damage with which to contend; there was also the cesium showering down onto the accumulated sludge. The situation was dire. In the district of Tamano, a mountainous area adjacent to the village of Iitate, a particularly large amount of cesium rained down onto the soil, contributing to a high ambient radiation dose compared with other areas. In addition to the damage wrought by the tsunami, therefore, we had a new problem: what to do about the damage caused by radionuclides.

What worried me most under these incredibly difficult circumstances was the thought that our farmers may lose the will to produce crops again. Nonetheless, I was able to find a ray of hope in the fact that the farmers still retained the strength of spirit to stand up to this catastrophe by farming as a community and forming agricultural corporations to work together.

It was around that time that the former president of Tokyo University of Agriculture, Kanju Ohsawa, visited us in Soma. I explained the situation to him at length, hoping that the university might perhaps be prepared to share its agricultural expertise to help us cope with the catastrophic circumstances we faced. He agreed to help, offering me the confidence-inspiring reassurance that the university would focus its assistance on Soma and work together with us to revive our agriculture. It was 2 months after the disasters when, buoyed by these kind words of support, we managed to rouse ourselves to collective action, determined to revive our agriculture one way or another. In addition to the assistance we received from the university, we were to rely a great deal on our community's strengths, as well as the desire of our farmers to start producing again. Here at city hall we provided leadership, acting as the overall coordinator to bring together the strengths of the university, the community, and the farmers as we pursued our goal.

Two years and nine months have passed since then, and thanks to the assistance we received from Professors Toshiyuki Monma, Itsuo Goto, and Yukio Shibuya, and the many other professors and students the university sent to help us, we are now achieving significant results in our efforts to revive our agriculture following the disasters.

One such accomplishment is in our cultivation of paddy field rice, where we have been able to harvest "Soma Revival Rice" after successfully using slag from steel-making to improve the quality of the soil impaired by the tsunami. In a separate initiative, our strawberry cultivation is benefiting from the efforts of strawberry farmers who joined forces to form an agricultural corporation and are working to deploy a new strategy to grow strawberries using hydroponics. Meanwhile, attempts to initiate measures to vertically integrate agricultural production, processing, and sales and distribution, are currently gaining momentum. In the district of Tamano, where the radiation dose was particularly high, a number of measures are proving effective, including farmland decontamination based on the results of soil surveys of individual parcels of land.

Now Tokyo University of Agriculture has compiled this book detailing all its initiatives in Soma to date. I very much hope that the book will serve to show as many people as possible that Soma is making steady progress with its recovery thanks to the university's agricultural expertise combined with our farmers' motivation and cooperative spirit and the strength of our community. We at city hall have also done all we can, but I think it is true to say that recovery can be achieved only when people involved in the various sectors affected receive substantial cooperation and assistance from members of their community and others. From that aspect, therefore, I am truly grateful for the assistance provided by everybody from Tokyo University of Agriculture. We still have much work ahead of us to revive our city, but I give my word that we will act together as a community to keep driving the recovery forward. As we do so, I sincerely hope that Tokyo University of Agriculture will continue to grant us the benefits of its agricultural expertise and the assistance of its teaching staff and students.

Mayor of Soma City   Hidekiyo Tachiya
Fukushima, Japan

Preface

The Tokyo University of Agriculture East Japan Assistance Project was launched in May 1, 2011, immediately after the occurrence of the Great East Japan Earthquake. By the time this book is published, the project will be in its fourth year of work. We still cannot forget the shock we felt on our first field survey, facing the paddy fields filled with rubble, the highly radioactive forests that no one was allowed to enter, the faces of people full of fear, and more. We stood, overwhelmed, with questions such as, "Can we really revive these disaster areas?", "Is there anything we can actually do to help?", and "Will the farmers be willing to resume farming?". However, we started the project with each project member determined to solve the problems one step at a time, to restore rich farmlands and bring back the farmers' smiling faces. This book is a record of our trial-and-error efforts and the fruit of our applied research, aimed at resolving the problems at hand.

Here, we summarize our latest results, in addition to those detailed in this book, to provide an overview of our achievements in this project. In our first challenge to recover paddy fields immensely damaged by the tsunami, we developed a convenient and low-cost farmland recovery technology called the Tokyo University of Agriculture Method, which has been adopted across Soma City. With this, rice has become producible on 267 ha of the damaged paddy fields in 2014, and an additional 200 ha will be recovered by the end of 2015. The first stage of recovery of paddy fields is almost complete. We also distributed the rice harvested in 2014 from the recovered paddy fields, named "Soma Revival Rice", to 4,000 elementary and junior high school students and teachers in Soma City. This activity has promoted local support for agricultural reconstruction, spreading attitudes such as "Let's eat these products together and support recovery!" and "Let's use them in school lunches!".

Regarding the second challenge of radioactive contamination problems, we have created a radioactivity monitoring system to cover each farm field in the Tamano area in Soma City, which is used to evaluate decontamination measures. In addition, we have performed experiments on radiation-contaminated grasslands of dairy farmers in Soma City to develop a decontamination method for steep-slope grazing lands. Once deemed too difficult to attempt and postponed by the Ministry of the

Environment, our research has attracted a great deal of attention. For the decontamination of trees with enormous amounts of radioactive contamination, we are developing together with local participants a new technology for the decontamination of persimmon trees and a technique for removal of cesium that has penetrated trees.

As can be seen, our efforts have produced results contributing to restoration, the first stage of reconstruction. However, there are still more severely damaged paddy fields that cannot be readily recovered from the effects of the tsunami. The development of methods to recover and farm on such paddy fields is an urgent issue we face. Other questions that remain in the agricultural reconstruction of tsunami-damaged areas include how to manage the finances and future business developments of agricultural cooperatives established after the earthquake.

Meanwhile, agricultural reconstruction of radiation-contaminated areas has just started, and we must study the means to support the future of farming here. Many of the contaminated areas are hilly and mountainous regions where agricultural production is difficult, providing little or no incentive for young people who left their home regions avoiding the radiation to return. In addition, damages caused by wild boar and monkeys have become more serious after the earthquake disaster, leading to more farmlands being abandoned and threatening the sustainability of rural agriculture. For the reconstruction of radiation-contaminated areas, we must create a new way of farming that will encourage young people to return to their home regions. It should be noted that overcoming reputational damage is crucial in the reconstruction of agriculture in Fukushima, and we must continue to address this issue. A lot of time is required for a complete recovery from the tsunami and radioactive damage of the Great East Japan Earthquake, and the Tokyo University of Agriculture will continue facing its challenges until this is achieved.

We would like to express our deepest gratitude to all the people who aided us in our project, especially the mayor, Mr. Hidekiyo Tachiya, and all other administrative staff of Soma City, the staff of the agricultural cooperative of Soma, the staff of Soso Norinjimusyo Fukushima Prefecture, and the farmers who supported our experiments. Also, we would like to extend our thanks to all students of the Tokyo University of Agriculture who volunteered to help in our field research and surveys. Our project would have failed halfway through without the support of all these people.

We close here with our heartfelt appreciation.

Tokyo, Japan	Toshiyuki Monma
Tokyo, Japan	Itsuo Goto
Tokyo, Japan	Takahisa Hayashi
Soma, Fukushima, Japan	Hidekiyo Tachiya
Tokyo, Japan	Kanju Ohsawa

Contents

Part I
The Road to Reconstruction from the Tsunami and Radioactive Contamination: Two and a Half Years On

Chapter 1
Dealing with Disasters of Unprecedented Magnitude: The Local Government's Tribulations and the Road to Reconstruction

Hidekiyo Tachiya

Abstract The large 9.0 magnitude earthquake, followed by the unexpected tsunami disaster, reached Soma City on the afternoon of March 11, 2011. In all, 5,027 citizens lost dwellings, and among them, 10 % lost their lives. The city urgently undertook major efforts in searching for dead bodies, rescuing survivors, and providing food and shelter for victims. In addition, the citizens of Soma were both extremely terrified and very confused by the severe radioactive pollution issues caused by the meltdown catastrophe at the 1st nuclear power plant of Fukushima. As the mayor of a city, I have conducted various measures to recover and reconstruct from such an unexpected disaster as quickly as possible. This essay tells the whole truth story about the trial and errors to overcome disaster. Also, I hope this essay will give you better understanding of the actual sufferings and efforts of the Soma citizens who cannot understand any other ordinary disaster records.

Keywords Soma city • Tsunami disaster • Radioactive pollution

1.1 Siege Conditions (March 24, 2011)

First, I wish to offer a prayer for the souls of the many people who lost their lives in the earthquake and tsunami.

Here in Soma, as soon as the earthquake's tremors stopped, I convened a meeting of the emergency response task force and instructed its members to issue a tsunami evacuation alert and guide residents to designated evacuation sites. Tragically, however, 5,027 people had their homes along the coast swept away and reduced to wreckage, and around one-tenth of that number lost their lives. To the volunteer firefighters who evacuated so many people I owe the deepest gratitude and respect, as well as my sincerest apologies. Seven firefighters gave their lives in the course of duty because it was their job to instruct residents to evacuate and lead them to

H. Tachiya (✉)
Mayor, Soma City Office, 13 Otesaki Nakamura, Soma, Fukushima 976-8601, Japan
e-mail: hisyo@city.soma.fukushima.jp

© The Author(s) 2015

T. Monma et al. (eds.), *Agricultural and Forestry Reconstruction After the Great East Japan Earthquake*, DOI 10.1007/978-4-431-55558-2_1

Fig. 1.1 The Haragama district, reduced to a mound of debris by the tsunami's onslaught

safety, and as a result they themselves were unable to escape in time. I feel that the only way I can try to atone for that loss is to do everything in my power to help rebuild the lives and community of the many citizens who were saved in return for those individuals' lives.

Immediately after the earthquake, we focused all our efforts on collecting information and rescuing survivors. At that point only one person had lost their life as a result of a building collapsing from the earthquake. But 50 min later the emergency response task force was informed of something unthinkable: the tsunami was about to cross the Route 6 bypass. I simply could not imagine it, but in fact the communities of Haragama and Isobe had already been obliterated, and Obama and Matsukawa had also been submerged by the waves, leaving only the elevated areas (Fig. 1.1). Nothing resembling a house was left. As anxiety and alarm coursed through our minds and bodies, the task force's next job was to protect survivors from harm and attend to the health of those who had been rescued. By evening the sea had engulfed everything along the coast, and all through the night we devoted ourselves to the task of finding as many as possible of those who had lost their homes and were alone amidst the waters. We took them to evacuation centers and gave them warmth, food, and water to drink.

From day two, the evacuation centers were overcrowded with those whose houses had been washed away, as well as others forced to leave their homes because of the lack of essential utilities. Somehow I think we just about managed to provide

Fig. 1.2 A tense atmosphere prevails at an emergency response meeting

the basics by distributing rice with the assistance of the female fire prevention volunteers and the self-defense forces, and by using the relief supplies that had already arrived. We worked together closely as a team, and began to take the very first steps toward implementing our medium- to long-term plans (Figs. 1.2 and 1.3). These measures would eventually include doing what we could to clear up the area affected by the earthquake and tsunami, helping people who had lost their homes to make the move from living in evacuation centers to living independently in apartments or temporary accommodation, and offering psychological care and help with staying healthy to citizens whose day-to-day activities would by then have been severely constrained for a long time.

But little did we realize that we would be assailed by a second "demon," this time from the far-flung county of Futaba, 45 km away (Fig. 1.4). This demon was the anxiety caused by fear of radiation. As the nuclear disaster escalated relentlessly, the frenzied reporting all day long incited terror, not only in people living near the power station but all across the country. From around the time the national government ordered residents within a 20-km radius of the power station to evacuate, the citizens of Soma started wanting to run far away, and that feeling began spreading.

At the same time, logistics service providers within Japan had become jittery and started to avoid entering the Soma area or the city of Iwaki. Gas tanker trucks and other delivery vehicles were stopping in Koriyama, so unless we sent our own drivers out we could not even obtain fuel. Meanwhile, the convenience stores and

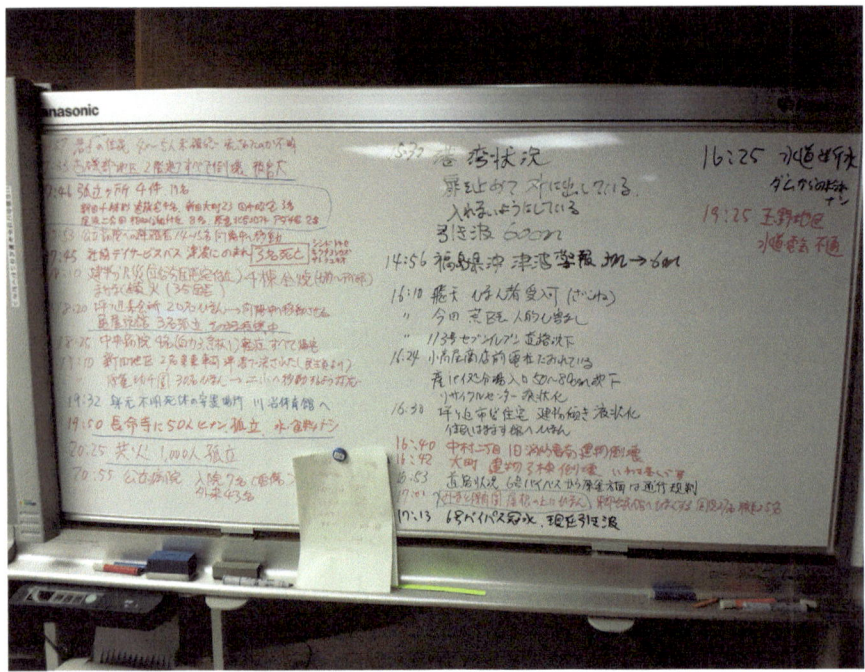

Fig. 1.3 A whiteboard lists issues to be resolved

Fig. 1.4 One of the steam explosions at the Fukushima Daiichi Nuclear Power Station that left people terror stricken about the radiation threat

supermarkets that had opened briefly after the earthquake no longer received any goods to sell, so they have closed. And in addition to the inconvenience of daily life with no supplies of gasoline or other commodities, fear of radiation spreading from the nuclear power station has swept through the Soma area. But it is out of the question for the emergency response task force to make an independent decision to voluntarily evacuate the city before the national government issues an evacuation order. That much seemed obvious, but we confirmed that it was in fact the case at an emergency response meeting, and have visited three evacuation centers to give speeches explaining the situation.

In emergency response meetings since then we have devised a plan of action to lead us along the path to reconstruction. We are distinguishing between short-term responses, medium-term responses, and long-term planning, and have decided to pursue our plan of action one step at a time according to the situation on the ground in Soma itself. If, in the course of that process, the national government instructs us to temporarily evacuate the city, we have to consider the health and lives of our citizens and implement a systematic mass evacuation. But if we allow a vague sense of anxiety to delay our plans for reconstruction, we will not be doing justice to the memory of those who died. My key concern is that life in evacuation centers outside Soma would be very hard on senior citizens and others who are particularly vulnerable when disaster strikes. I therefore have no intention whatsoever of taking Soma's citizens away from the city at this stage, when no evacuation order has been received from the national government.

The commotion surrounding the nuclear disaster and the situational damage affecting the distribution of commodities are unlikely to continue forever. If we give in to them, we will be letting down the memories of the volunteer fire chief and firefighters who gave their lives while on duty protecting the residents of our worst-hit communities from the tsunami. We can survive so long as we at least have rice, miso, and pickled plums. The Tenmei famine that our ancestors endured in the 1780s must have been worse than this. So we will persevere here under siege conditions. Fortunately, the mayors of other cities across the country are behind us, so we do not need to worry about our food supply running out.

1.2 Mayoral Address to City Government Employees (March 22, 2011)

Today is the twelfth day since the earthquake and tsunami. First and foremost I would like to express my deepest appreciation to all of you for your calm, reliable, and orderly response under the emergency response task force's chain of command during these past days.

Immediately after the earthquake struck we set up the emergency response task force to make the major decisions about how to proceed. The task force decided that we should first of all protect citizens living along the coast from any tsunami triggered by the earthquake, and that we should rescue any people who might still be alive in houses that had collapsed. However, the actual tsunami was huge—far bigger than we could have imagined. That was what made this particular disaster different from any other.

8 H. Tachiya

Fig. 1.5 Saving lives by rescuing survivors and opening evacuation centers promptly

Once the tsunami hit, our primary concern was naturally to safeguard lives amidst this unprecedented natural calamity. Our response in the immediate short term, therefore, was to save people waiting to be rescued, and to evacuate people from dangerous locations to places where they would be safe. Of course, searching for missing people in the course of that process has also been part of the short-term response. We have now reached the stage where the likelihood of being able to save any more survivors is close to zero. But the fact is that there are citizens like you and me who lost their lives and are now still lying under the debris. That is something we need to deal with properly going forward (Fig. 1.5).

But what are the medium- and long-term issues we also need to consider? In the medium term, we need to help people living in evacuation centers to move within the city to apartments, disused public housing, or newly built temporary accommodation, where they can lead independent lives. In the long term, meanwhile, our ultimate goal is of course to ensure that reconstruction is achieved in the districts where the impact of the disaster was particularly catastrophic, such as Haragama, Matsukawa, and Iwanoko, as well as Isobe.

Moving on to the nuclear issue, our citizens are now experiencing considerable anxiety in this regard, and I would like to consider it in terms of two distinct goals. One is ensuring that our citizens suffer no damage to their health as a result of radiation from the power station; the other is to ensure that they do not end up changing the way they live their day-to-day lives because of anxiety about radiation.

I am 99 % certain that the government will not issue Soma with an evacuation order, but the possibility cannot be completely denied, so I believe we need to anticipate that eventuality as well, and I am in fact thinking about what would happen in that case. At the initial planning stage our fundamental policy was to take a slow and steady approach in deciding both how to reconstruct our city—a substantial area of which has been destroyed by the earthquake and tsunami—and how to maintain quality of life for our citizens during that process. That is the course that Soma is following. The nuclear problem does not change that in any way. We will still continue to plant one foot firmly in front of the other to follow our chosen course. Unless the national government issues an evacuation order, I intend to continue moving forward one step at a time in line with our initial reconstruction plan.

That is the course I will be pursuing, so I would like to ask you all please to work together and help me to revive Soma. And I would also ask you to encourage our citizens to do the same. Thank you.

1.3 Soma's Reconstruction Plan (June 12, 2011)

On June 3 we launched the Soma Reconstruction Council to deliberate on the future course of reconstruction in Soma. I listed the issues requiring discussion as I saw them and presented them as points for consideration. But at this stage it was still not really possible to draw a clear picture of what the disaster zone will look like in 3 or 5 years, or of the kind of life those affected by the disasters see themselves living. Part of the problem is that progress will be impossible unless systemic changes are made at the national level, but more importantly, our plans will remain mere pipe dreams unless we have some prospect of obtaining the necessary funds. While heading the emergency response task force during the past 3 months I have placed top priority on minimizing the harm caused to people's physical and mental well-being by the disaster, and on preventing communities from falling apart as a result of the nuclear crisis or other sources of secondary damage. Everyone at City Hall, and the citizens in general, have pulled together and persevered to promptly and carefully handle each new problem as it arises, and consequently we were able to avoid anybody dying as a consequence of delayed action. The members of the emergency response task force have done a particularly good job, working with virtually no time off at all.

Immediately after the disasters, the national government launched its council for reconstruction, and to be honest I felt uneasy about it at the time. For at least the first month after the earthquake we were completely absorbed in dealing with immediate problems, and the issues they were discussing at the council seemed completely foreign to us. These disasters were on an unprecedented scale and affected every municipality in the area in a distinct way. The big difference compared with other disasters is that reconstruction is not going to be a case of simply returning to what we had before. Even the definition of what "reconstruction" means varies according to the unique nature of each community. Municipalities in Fukushima Prefecture have all been affected by the nuclear accident to a greater or lesser extent, and their process of reconstruction is likely to take a completely different form from the reconstruction process of municipalities in Iwate and Miyagi Prefectures. Even within Fukushima, the extent of the damage and the attitudes toward reconstruction differ from one municipality to another. Only the citizens of Soma can possibly understand Soma's situation, so we have no choice but to rack our own brains to work out a reconstruction plan and nurture it into being. I think the government's reconstruction council should liaise closely with the municipalities affected by the disasters and discuss measures for reconstruction based on the realities of their situations. What distinguishes the colossal damage sustained from a huge tsunami such as this is the fact that there is no way reconstruction can be achieved by returning to the way things were before it happened (Fig. 1.6).

Fig. 1.6 The Soma Reconstruction Council explores reconstruction measures appropriate to the local situation

Another key consideration is the need to manage the long journey to reconstruction. However impressive our reconstruction plan may be, it would be tragic if anybody died alone or took their own life along the way. That is why our reconstruction plan includes a strategy for managing the entire process until our goal is finally achieved. Naturally it also includes measures to address the many challenges we anticipate, including maintaining the health of people living in temporary accommodation, preventing people from dying alone, caring for children suffering from posttraumatic stress disorder (PTSD), protecting the health of people working to remove debris, revitalizing the local economy, and solving problems related to radiation. Resolving these issues is essential to our reconstruction, so the central concern will naturally be to rebuild the lives of the citizens affected by the disasters. At the risk of overstating our aims, I personally think that reconstruction could best be defined as a situation in which those affected by the disaster, whatever their age, are able to make new life plans.

1.4   Haragama Morning Market (August 18, 2011)

Haragama, where I was born and raised, has been a fishing village since time immemorial. When I was a boy the fishermen would go fishing in the rowing boats that were lined up on the shore, but from about the 1960s onward motorized boats took over and the coastal fishing industry, the port, and the associated industries all steadily increased in size. As a result, the fiscal 2010 catch at Matsukawaura fishing port was valued at slightly under 5 billion yen. At the same time, related businesses such as wholesaling and, in the case of certain types of fish, processing, also developed.

The tsunami was so massive that more than half of the 300 or so coastal fishing boats disappeared, but about 40 % of the fishermen made a dash for their boats when the earthquake hit and drove their engines to the limit heading for the open sea where the tsunami was looming. Even when a massive tsunami threatens, if the fishermen can just manage to get past it before the waves break, they can wait safely out at sea until it is over. But if they get caught in a wave that has broken, their boat will be smashed to smithereens. In fact, there were some fishing boats that were engulfed by this enormous tsunami because they were only very slightly behind the others that survived. Afterward, the fishermen who had been out all night in the open sea had difficulty reaching the shore when they returned because the port's quay had been damaged, but once they ran out of food and water out at sea they had no choice but to attempt the risky return to port. One can hardly imagine what they must have thought when they finally managed to reach dry land and witnessed the sight of Haragama changed beyond all recognition.

Then, having lost not only their homes, but also family members and relatives, they were relentlessly assailed with demands for repayment of loans on their fishing boats and equipment. One particularly conscientious fisherman begged me to get the fishing port at least temporarily restored as soon as possible, saying that he wanted to go fishing again as soon as he could, because if he did not ultimately he would have to commit suicide because of the loans he owed. Another fisherman had resigned himself to not taking his boat out into the open sea when the tsunami was looming, and instead escaped by car to drive his grandchildren to safety. He chose his grandchildren's safety knowing that he would most likely be plagued by a loan repayment nightmare afterward. A fisherman who loses his boat has to invest several tens of millions of yen to continue making a living, whether he buys a second-hand or a new boat. In most cases fishermen have not finished paying the loan on their previous boats, so any new expenditure entails so-called "double loans" (having to take out a new loan to replace an asset for which the existing loan is still outstanding). Attempts to restart the fishing industry are now hindered by the problem of these double loans. That is why we are very keen to see the enactment of the double loan relief bill currently under consideration in the Japanese national assembly. I sincerely hope that somehow it can be passed soon.

One day in May some wholesalers who still remained locally came to talk to me about how they might raise spirits in Soma despite the devastation of its fishing industry by delivering seafood to our citizens. I suggested that they could try organizing a morning market similar to the famous one held at Wajima in Ishikawa Prefecture. These people had good intuition, and they came up with a whole succession of new ideas. They suggested forming an NPO (nonprofit organization) to provide a continuing supply of cooking ingredients, and eventually creating a major food market in Haragama that would be used by citizens as their main source of

Fig. 1.7 The morning market acts as a spur to both morale and industrial reconstruction

food. They started to put the plans into action immediately, and since then 3 months have passed. The morning market, started on the Nagatomo Sports Ground in front of Soma Nakamura shrine, now attracts 2,000 citizens a day.

When you start from zero, nothing happens if you do not chase your dreams. But in this case something is already happening. The NPO was approved by the prefectural authorities, and is currently in the process of being registered. Meanwhile, the morning market, now held on weekends in the foyer of the Soma City Sports Arena, is doing a roaring trade and has become something for citizens to anticipate. The Haragama Morning Market NPO took action to revitalize our community, and our hopes for the disaster zone and for ourselves are invested in it (Fig. 1.7). Nuclear contamination and other major obstacles stand before us, but we will keep marching onward and upward!

1.5   Assistance from Tokyo University of Agriculture (March 11, 2013)

The day after the earthquake and tsunami I went to see the inundated rice paddies in the districts of Iwanoko, Niida, Kashiwazaki, Nittaki, Isobe, Yunuki, Kita Haragama, and Niinuma. I was lost for words. The roots of pine trees carried from the Osu district reached upward, exposed, as if they were growing on top of the sludge. Scattered among them was the debris of wrecked houses.

Toward the end of April that year, when the commotion surrounding the nuclear disaster had calmed down to some extent, we started to hear some pragmatic points of view from the farmers affected by the disasters. They said that they could not give up the idea of farming because at their advanced age there was nothing else they could do, so naturally they wanted to farm again. But the cost of buying new farming machinery was prohibitive, so they could not return to farming unless it was possible to take an efficient, community-based approach. When I heard that,

I thought that perhaps these first stirrings of motivation could be cultivated into renewed hope for the future. Community-based farming can be described as the first step toward forming an agricultural corporation, and now that even agriculture is exposed to the effects of globalization, it seems logical that farmers should seek to farm more efficiently. Rather than small-scale farmers having to make excessive individual investments in machinery such as tractors, therefore, they could integrate their farmland through their own business corporation to farm on a larger scale. Hearing what the farmers said, we sprang into action to organize them by establishing agricultural corporations. We did not have to involve all the farmers initially; so long we could create a few successful corporations, then other farmers were likely to follow suit. I had a series of spirited discussions with the chief of the Tohoku Regional Agricultural Administration Office and he promised that he would provide technical assistance to set up the agricultural corporations, and as much support as possible thereafter.

Then, one day in May, I had a visit from Dr. Kanju Ohsawa, the president of the Tokyo University of Agriculture, whose students were helping us as volunteers at the time. Drawn by his unassuming manner, I opened up to him and asked his opinion about the action we had taken since the disasters, and the challenges we still faced. And he responded by saying that, if we wished, Tokyo University of Agriculture would do all it could to help Soma in its efforts to recover. At that point, our major challenges involved finding ways to (a) restore the disaster-stricken farmland covered in washed-up trees and sludge and use it to generate income for the individual farmers making up the agricultural corporations; (b) decontaminate the land and revive agriculture in the district of Tamano (the community adjacent to the village of Iitate), where the radiation dose was relatively high; (c) revive the strawberry-growing association in Wada, half of whose members sustained damage as a result of the disasters, losing their production base; (d) take advantage of this opportunity to consolidate some of the city's farmland, transforming it into new, larger paddy fields; and (e) find new uses for sites offering no prospect of future use as farmland because the farmers responsible had abandoned them as excessive debris made restoration impossible. I asked Dr. Ohsawa for technical guidance in handling these challenges, and we agreed that he would join the Soma Reconstruction Advisory Board, to be launched in June.

In the summer of 2011, Professors Toshiyuki Monma, Itsuo Goto, and Yukio Shibuya came to survey the sites and give practical guidance with the logistical support of Dr. Katsumi Takano, the vice-president of the university. At City Hall we embarked on setting up agricultural corporations and started telling farmers about them with a view to persuading the more motivated farmers to organize themselves. We also introduced the academics from Tokyo University of Agriculture to local people, explaining what they were doing so that they would be accepted within the communities. At first it was a little awkward, but I think that over time they earned the residents' trust. Professor Monma, who was responsible for radiation countermeasures in Tamano, rented an apartment and lived permanently in Soma while he was working here.

Fig. 1.8 Students from Tokyo University of Agriculture measure radiation in a paddy field

The staff and students from the university surveyed the contaminated soil, taking detailed measurements from one parcel of farmland at a time. Their survey entailed measuring the ambient radiation dose at two different heights and sampling soil from two different depths. They then explained the results of their survey to residents (Fig. 1.8). Their enthusiasm for their work was such that the residents of Tamano would follow Professor Monma's advice, even if they would not listen to what I said. Because of the high ambient radiation doses and the reputational damage, we refrained from planting any rice in 2012, but took the opportunity to reduce the soil radiation dose by planting green manure crops and working them into the soil using deep plowing. As a result, we are planning to plant rice in 2013, and Professor Monma and his colleagues are now thinking about ways to revive our vegetable and dairy farming.

Meanwhile, in the area including the districts of Iwanoko, Niida, and Kashiwazaki, where the village of Iitoyo used to be, soil expert Professor Goto came in to lead the reconstruction effort. I thought that, even if we managed to clear away the washed-up trees scattered everywhere, it would be a huge challenge to restore paddy fields on which sludge had accumulated several tens of centimeters thick. However, right from the beginning Professor Goto said something puzzling: nothing could beat rain for removing salt, and reversing the soil with a plow would turn the sludge into fertilizer. So in other words, if we removed all the trees and debris and just furrowed the fields, the salt would disappear naturally. However, we would also have to deal with sulfuric acid produced by the substances the sludge had brought with it, and Professor Goto suggested that we do this by using steelmaking slag to rebalance the

Fig. 1.9 Yamato Welfare Foundation's donation of farming machinery spurred the establishment of agricultural corporations

pH value of the soil and replenish minerals. Naturally, we were surprised at the idea of using a steel by-product as fertilizer, but the professor said he was convinced it would work, and as we listened to him we were all eventually persuaded. Then Nippon Steel & Sumitomo Metal Corporation offered us the slag we needed, and thanks to the kindness of Keiji Aritomi, the president of Yamato Welfare Foundation, we also received a donation of tractors and other farming equipment to replace those that had been washed away in the tsunami (Fig. 1.9). In the fall of 2012, the "plowed-sludge steelmaking-slag-enriched rice" cultivated by the Iitoyo District's newly launched agricultural corporation achieved an impressively bumper harvest. It was only 1.7 ha of paddy fields that had been restored, but it was like a dream come true considering the hopelessness we had felt the day after the earthquake and tsunami (Fig. 1.10). No doubt my feelings at the time had something to do with it, but that steelmaking slag rice was the best I had ever tasted in my life.

On March 8, 2013, we held a presentation ceremony, formally accepting donation of a large quantity of steelmaking slag from Nippon Steel & Sumitomo Metal Corporation to help us in our goal of planting 50 ha of rice during the April 2013 to March 2014 fiscal year. The ceremony was held in Soma's City Hall with the university's vice-president, Dr. Katsumi Takano, in attendance. As my name for the rice was not really an option, the 1-kg bags on display were labeled "Revival Rice," but we need to think of a more sophisticated name, like the title of a Japanese *enka* ballad, that is evocative of the hardships we have overcome. On a different note, the soil decontamination and salt removal process, combined with the high concentration of potassium in the sludge, make any cesium contamination unlikely, but we need to be considerate toward the purchasers of our rice and test every bag for cesium anyway.

In addition to the rice-growing corporation, the strawberry-growing association also formed an agricultural corporation and has been very successful. But if we are to build on such successes to achieve agricultural revival, we need to make decisions about how land is to be used, and we cannot do that unless basic surveys and research are conducted repeatedly. For these reasons, I would like to continue working with Tokyo University of Agriculture on an ongoing basis. I am deeply grateful to our

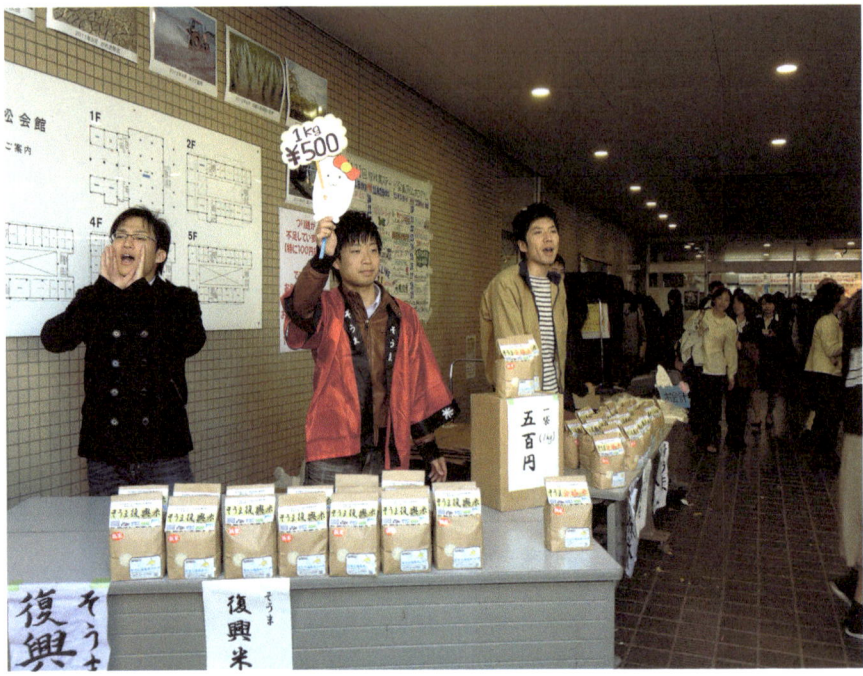

Fig. 1.10 Students at Tokyo University of Agriculture sell Soma Revival Rice

partners at the university for the kindness they have already shown toward our city. We, for our part, intend to work hard to accomplish an agricultural revival that will earn worldwide recognition and be appreciated by future generations.

1.6   The Wada Strawberry Farm Agricultural Corporation (March 11, 2013)

The Wada pick-your-own strawberry farm was doing a roaring trade until just before the earthquake. Cheerful staff welcomed visitors to the greenhouses of the 13 strawberry-growing association members involved. It was very popular with children, and particularly kindergarten pupils. Since it opened in 1988, the strawberry farm's plastic greenhouses had become a seasonal fixture in Soma between January and May, welcoming 30,000 to 50,000 visitors in an average year. However, the tsunami swept away about half of these in an instant, and the surrounding paddy fields were left unrecognizable because of the thick layer of sludge brought with the tsunami. Then, the stunned strawberry farmers suffered a further blow, this time as a result of the nuclear disaster. Admittedly, the strawberries were grown inside greenhouses, so even if the plastic had been washed away and cesium had attached itself to the soil, provided the greenhouses were rebuilt and soil was brought in from elsewhere, strawberries grown there subsequently would not be contaminated by radiation. But, given the reputational damage caused by the radiation, how many strawberries could actually be sold? If fears of radiation caused even logistics service providers to give the region a wide berth, would any visitors actually come? Perhaps because they could see that I was worried, the older association members who had lost their greenhouses in the disasters started to give up on the idea of investing to rebuild them. Of course, growing strawberries takes considerable time and effort, and it is work that involves bending down over the ground, which is hard for older people, so I felt that it would be wrong of me to just blithely try to drum up enthusiasm. But the head of the strawberry-growing association, Kenichiro Yamanaka, had not given up hope. He enlisted the help of a group of students from Tokyo University of Agriculture and other volunteers from all over the country, and began the process of preparing to reopen the strawberry farm. During that time immediately after the disasters it was only our motivation to recover that provided a ray of hope, and he was very strongly motivated to reopen the strawberry farm, even if it was only with the few strawberry farmers who had escaped the ravages of the tsunami.

It was at that point that a young man called Hirasawa, who was the son-in-law of the former chairman of the medical association, visited me with a proposal. He said that it was technically possible to create round, dome-shaped greenhouses supported by positive air pressure, and he suggested the idea of using them to grow strawberries hydroponically (in nutrient-enriched water instead of soil). It made a lot of sense to me: hydroponic cultivation would save the farmers from back pain, and they could put systems in place to fertilize and monitor the water. But if we implemented this scheme as a reconstruction project, focusing solely on the farmers who had lost their greenhouses to the disasters, it would only create a gulf within the association. So should we not instead get all the association members to participate in creating an agricultural corporation, then we could erect a big hydroponic greenhouse for strawberries as one of the city's reconstruction projects and lease it from the city to the corporation free of charge. Having hit on this idea, I felt compelled to suggest it to the members of the strawberry-growing association. However, the suggestion of an agricultural corporation was new to some people, and they did not understand what I was talking about. In the end it proved impossible to persuade everybody to join, but we decided instead to offer shareholdings in the corporation to those who felt they were simply too old to actually grow strawberries any more. The head of the city's industrial affairs department visited such association members repeatedly to secure their support.

Thus, we set up the Wada Strawberry Farm corporation and constructed our first giant plastic greenhouse. Then, on January 13, 2013, we were joined by the national government's newly appointed Minister for Reconstruction, Takumi Nemoto, for the ceremonial opening of the new pick-your-own strawberry farm. The association's members were grinning from ear to ear that day. The minister was an old acquaintance of mine, so we were happy to see each other, but I was especially pleased to be able to show him how an agricultural corporation could operate a restored farm that enabled strawberry farmers to focus on agriculture and tourism without having to worry about radiation. That afternoon, Yoshimasa Hayashi, the Minister of Agriculture, Forestry and Fisheries, also came to inspect the farm.

Fig. 1.11 Wada Strawberry Farm, the symbol of Soma's agricultural reconstruction

Unfortunately we ended up not using Hirasawa's round, domed greenhouses, but it was my encounter with him that switched a light on in my mind, and for that I am most grateful (Fig. 1.11).

Chapter 2
Tokyo University of Agriculture East Japan Assistance Project Assisting with Reconstruction: Guiding Principles, Planning, and Propagation of Benefits

Toshiyuki Monma

Abstract The Tokyo University of Agriculture East Japan Assistance project was started in April 2011 with Soma City, Fukushima Prefecture, as a target area. Three years having already elapsed since the start of the project, in the tsunami-affected areas, we contributed reconstruction in the tsunami damage paddy field by development and dissemination of restoration technology by the Tokyo University of Agriculture method and support of new farming organization construction. In addition, in the radioactive contamination area, we contributed clarification of the actual situation of radioactive contamination, and the decontamination method of the forest, and could restart farming by development of the radioactivity monitoring system.

Keywords Tokyo University of Agriculture • East Japan Assistance project • Tsunami damage • Radioactive contamination

2.1 The Launch of the East Japan Assistance Project

2.1.1 Initiatives to Support Disaster Zones and Students Affected by the Disasters

The Great East Japan Earthquake struck 18.1 s after 2:46 p.m. on March 11, 2011, and it ranks among the most devastating disasters ever recorded anywhere in the world. The following morning, at Tokyo University of Agriculture, we focused all our efforts on checking for earthquake damage to our classrooms, laboratories, and other facilities, and on establishing which of our students had been affected by the

T. Monma (✉)
Department of International Biobusiness Studies, Tokyo University of Agriculture,
1-1-1 Sakuragaoka, Setagaya-ku, Tokyo 156-8502, Japan
e-mail: monma@nodai.ac.jp

earthquake and tsunami. We moved as quickly as we could to gather information about how our students had been affected, covering each faculty, department, and research group one by one. By March 31, we had completed our final checks on the safety of students whose homes were in the areas designated by the government as disaster zones (spread across the prefectures of Aomori, Iwate, Miyagi, Fukushima, Niigata, Ibaraki, Tochigi, Chiba, and Nagano). We established that 9 of our students had parents who were missing, 5 had other family members who were missing, 12 were in evacuation centers, 7 had lost their homes, and 21 had homes that were damaged.

We also considered the need for concrete measures to provide financial aid to those who required it among the 335 students from the government-designated disaster zones scheduled to start at the university in April, or among the 1,401 students from the disaster zones already enrolled at the university, and identified 204 students requiring aid. Specific aid measures included reducing or waiving tuition fees and subsidizing living expenses. Actual financial aid provided came to approximately 200 million yen in the academic year from April 2011 and 110 million yen in the academic year from April 2012.

2.1.2 Our Desire to Help the Regions That Had Sustained Tsunami Damage and Radioactive Contamination, and the Launch of the Project

Once the plans for providing aid to students had taken shape, we became aware of a rising groundswell of opinion among teaching staff and students who were saying that we should do something on behalf of the university to assist people in the disaster zones, or that they personally wanted to do what they could to help. The students in particular were saying that they wanted to work as volunteers to help people in the disaster-stricken areas. Meanwhile, teaching staff were saying that we as a university should do something to help revive the agriculture, forestry, and fishery industries in those areas, both to help with restoration and recovery and for the sake of the many students at the university who had suffered as a result of the disasters. Hearing this, the university authorities promptly embarked on a project to offer practical assistance for recovery in the disaster zones by drawing on the body of agricultural research amassed by the university. Overall guidance was provided by the former president of the university, Kanju Ohsawa (currently chairman of the board for the educational corporation that operates Tokyo University of Agriculture and its associated schools), and I was appointed project leader. I was in hospital at the time, but while I was confined to my sickbed I worked out what the project would entail, the organizational structure for undertaking research, and the fields to be researched. I then drew up an overall strategy for providing research-based assistance in the disaster zones and submitted it all to the university authorities. The project plan was deliberated first by the council of faculty heads, then by the university council, as a result of which an annual budget of 18 million yen was secured from internal funds and a decision was made to put the plan into action. The project was named the Tokyo University of Agriculture East Japan Assistance Project.

2.2   The East Japan Assistance Project Gets Under Way

2.2.1   Why Soma Was Chosen for the Project

There were a number of reasons why Tokyo University of Agriculture decided to start its postdisaster reconstruction project in the city of Soma. The Great East Japan Earthquake was a multiple disaster without precedent: in addition to an earthquake and tsunami, phenomena the Japanese people had experienced before, a new, completely unheard-of disaster was added in the form of radioactive contamination, and that in turn triggered further problems from reputational damage. The key factor in our choice of location, therefore, was our shared understanding that to achieve a genuine recovery following the Great East Japan Earthquake it would be essential to develop and promote technologies and methods designed to address four separate issues: the earthquake, the tsunami, the radioactive contamination, and the reputational damage.

Furthermore, the Tokyo University of Agriculture's educational philosophy is founded on a practical-science-based ethos, the essence of which is expressed in the words of the university's first president, Tokiyoshi Yokoi, who exhorted students of agriculture to "Ask the rice plant about the rice plant; ask farmers about agriculture." Adhering to this philosophy would therefore mean identifying the issues on site, developing scientific technologies and methods to resolve those issues, and conducting trials to demonstrate the efficacy of the technologies and methods, again on site. Our second shared understanding, therefore, was our commitment to maintaining a practical-science-based ethos from beginning to end as we undertook the East Japan Assistance Project.

We then added two more conditions to the factors already detailed: (a) our target disaster zone should not be closed to public access; and (b) there should be no prohibitions on planting crops, etc.

Having reduced the number of potential target sites for the assistance project by means of the foregoing conditions, we were left with the cities of Soma and Minamisoma as our two final candidate locations. Right up until the last minute we were torn between these two, but in the end we decided on Soma because it had no restrictions on planting paddy rice.

2.2.2 The First Comprehensive Surveys and the Initial Shock

In getting the East Japan Assistance Project off the ground, the first major issue was how to forge a common understanding among the project team members about what they needed to do, what it was possible to do, and how it would be done. So we followed the maxim "seeing is believing," and visited Soma from May 1 to 4 to fully experience the extent of the damage there for ourselves. Once there, we held meetings with the Fukushima Agricultural Technology Centre's Hama District Research Institute and the city authority to discuss how the project would be implemented, and conducted surveys of the area. The group comprised the former president of Tokyo University of Agriculture, Kanju Ohsawa (who was still president at the time), myself as project leader, 12 members of the teaching staff, and two researchers. While conducting our on-site surveys we reached agreement with the relevant organizations on the details of how we would cooperate with them to implement the assistance project.

2.2.2.1 Research Collaboration with Fukushima Prefecture

A meeting was held at the Fukushima Agricultural Technology Centre's Hama District Research Institute in Soma to discuss our cooperation with Fukushima as a prefecture. It was confirmed that we would cooperate with the Hama District Research Institute on research to develop technologies to overcome the damage caused to fields by the incursion of seawater, earth, and sand. Meanwhile, we would also cooperate with the Fukushima Agricultural Technology Centre itself on research to ascertain the extent of damage to the region's farming and agricultural production structures and to investigate measures to provide assistance.

2.2.2.2 Soma's Challenges and the University's Initiatives

In our meeting with Soma's city authority, the mayor, Hidekiyo Tachiya, asked for our help with recovering the following agricultural issues: (a) identifying which of the approximately 1,100 ha of paddy fields covered in sludge and debris by the tsunami could be recovered and used again, and which could not; (b) investigating what specific measures should be taken to revive agriculture and farming villages; (c) investigating measures to deal with reputational damage now that the adverse effects resulting from radioactive contamination were lasting longer and becoming more serious than initially anticipated.

2.2.2.3 Shock During Our On-Site Surveys

Our on-site surveys involved inspecting the districts that had suffered particularly serious damage, guided by local residents who were assisting with the project. We had all to some extent formed our own impression of the aftermath of the disasters

Fig. 2.1 Dumbfounded at the utter devastation wrought by the tsunami

from what we had seen and heard on the television and in newspapers. Nonetheless, when we saw with our own eyes what had actually happened there, the utter devastation wrought by the tsunami left us all dumbfounded. We wondered whether these districts really could be revived, whether there was actually anything we could do to help, whether the farmers would ever be able to recover from this (Fig. 2.1). But, shocked as they were, the teaching staff immediately started collecting soil samples, investigating how the disaster-stricken communities had been affected, and interviewing farmers about the damage to their farms. Thus, somehow, the surveys in each field of specialization came to be completed. These on-site surveys shed light on various issues causing concern as well as problems requiring solutions, and afterward each member of staff was left to think about what could be done in his or her own specialist field and then to put ideas into practice. The university's practical-science-based ideology calls on us always to prioritize solving problems at the location where they occur, and with this approach we were well equipped to work toward postdisaster recovery.

2.2.3 Project Structure

Based on the on-site surveys, we agreed that during the course of the East Japan Assistance Project our policy would be to act to resolve issues in response to requests for help from the disaster zones, rather than our researchers acting according to their own research interests. We structured the project as shown in Fig. 2.2 to enable us to respond to whatever requests we may receive from the areas affected by the disasters.

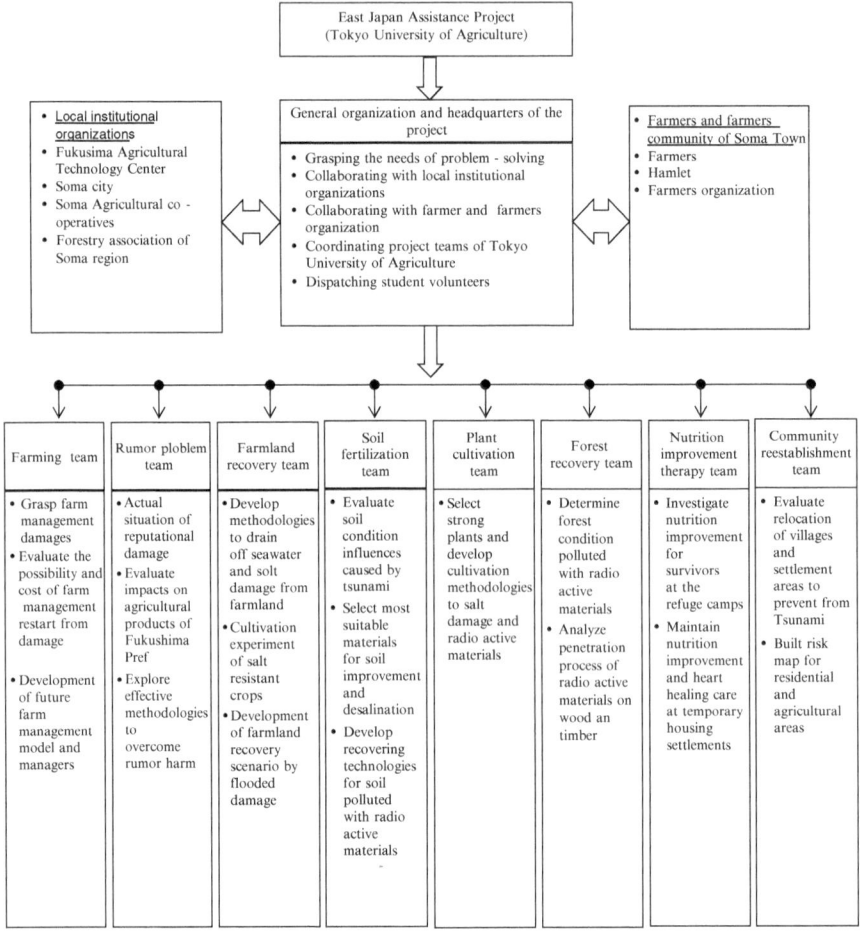

Fig. 2.2 Structure of the Tokyo University of Agriculture East Japan Assistance Project

2.3   Start of Initiatives and Initial Research Outcomes

2.3.1   Student Volunteers

Although they are not included in the project structure shown in Fig. 2.2, our student volunteers have played an extremely important role in the smooth implementation of our on-site initiatives and in establishing a framework for cooperating with the farmers affected by the disasters. While implementing the project we have responded to the farmers' requests, recruiting student volunteers specifically to help with reviving agriculture and sending them to the city of Soma for approximately 1 week, which is longer than most volunteers stay. The farmers asked us to send

Fig. 2.3 Greenhouses recovered with the students' help (Photographs taken June 29, 2011)

Fig. 2.4 Students earned the sincere gratitude of residents during their week in Soma

student volunteers to help them with a number of tasks that included recovering the strawberry greenhouses damaged by the tsunami, helping to grow strawberries so they could restart strawberry production, and undertaking crop thinning to help the *nashi* (Asian pear) farmers who were unable to manage the work alone because they had lost their homes and farming machinery in the tsunami. The university's international students have also taken part in these voluntary activities, working up a sweat through hard work in the fields, and making a major contribution to speeding up the recovery (Figs. 2.3 and 2.4). The student volunteers' efforts have become well known among Soma's farmers, and were instrumental in helping them to appreciate that the Tokyo University of Agriculture's project was not just another transitory research venture.

It would be fair to say that student volunteer activities such as these were the main reason why the university's project to support recovery in the city of Soma was warmly received by the farmers.

2.3.2   Farming Team

After the project was launched on May 1, 2011, the farming team undertook activities broadly divided into three types, as follows:

1. Ascertaining the extent of agricultural damage and assessing desire to resume farming
2. Investigating factors determining motivation to resume farming and measures to assist farmers in starting again
3. Assessing the activities of existing agricultural reconstruction associations and investigating reconstruction models

The farming team addressed these issues mainly by researching the activities of farmers and communities affected by the disasters, as well as the activities of agricultural recovery associations. The research produced the following key outcomes with regard to the revival of farming (for further details, refer to Chap. 5).

2.3.2.1   Investigating the Extent of Agricultural Damage and the Preconditions for Resumption of Farming

The team demonstrated that the following measures would be important in enabling key farmers to resume their farming activities: (a) checking for radionuclides; (b) obtaining an assurance from the national government or Tokyo Electric Power Company (TEPCO) that they would compensate farmers by buying agricultural produce if radionuclides were detected and produce could not be sold; (c) recovering farmland, irrigation facilities, and other key infrastructure promptly; and (d) establishing frameworks for cooperating to achieve recovery in the various communities. The team's research also indicated that measures implemented by individual communities would play an important role in postdisaster agricultural revival.

Immediately after the disasters, many of the farmers showed a strong desire to restore their farms to the way they were; it was clear, therefore, that there was little prospect of introducing new types of farming or new crops. Furthermore, there were two factors of major importance in determining whether those farmers who had suffered tsunami damage wanted to return to farming: (1) the extent of tsunami damage to their farmland; and (2) the extent of tsunami damage to their farming machinery. The team identified the following four patterns (see Fig. 2.5) that characterized farmers' degree of motivation to resume farming according to the combination of major or minor damage to their land and machinery (Monma 2013).

Pattern 1: Desire to Resume Farming Weak: Both land and machinery sustained serious tsunami damage, and the farmer cannot restore them on his/her own. Key preconditions for recovery are reconstruction of farmland, irrigation facilities, and other infrastructure by the national government, as well as creation of a structure to share farming machinery within each community. Aid from the national government is also necessary.

	Damage of agricultural machinery: serious	Damage of agricultural machinery: slight
Tsunami damage of farmland: serious	Desire to resume farming: weak	Desire to resume farming: strong
Tsunami damage of farmland: slight	Resumption of farming: dependent on preconditions	Desire to resume farming: strong

Fig. 2.5 Level of desire and preconditioning for resumption of farming among farmers who suffered tsunami damage

Pattern 2: Desire to Resume Farming Strong: Despite serious tsunami damage to the farmer's land, his/her machinery was undamaged. In this case, provided that the national government recovers farmland, irrigation facilities, and other infrastructure, the farmer can resume farming on his/her own thereafter.

Pattern 3: Resumption of Farming Dependent on Preconditions: Even if the tsunami damage to their land was relatively slight, farmers whose machinery was destroyed or damaged are unable to replace the machinery on their own. Aid from the national government for purchasing new farming machinery is therefore essential. The form that subsequent farming takes will depend greatly on whether the government requires the farmers concerned to create an organization to share farming machinery or whether individual aid will be provided.

Pattern 4: Desire to Resume Farming Strong: Farmers who suffered minimal tsunami damage to their land or machinery demonstrated a desire to resume farming promptly. Provided that irrigation facilities and other infrastructure were not damaged, the individual farmer could deal with such tasks as removing the salt from his/her land, but if radionuclides had accumulated, then decontamination would be essential.

2.3.2.2 Agricultural Recovery Association Activities: Characteristics and Challenges

Purpose of the Research and Survey Method

Japan's Ministry of Agriculture, Forestry and Fisheries made it a policy objective to reconstruct local agriculture in the disaster-stricken areas and resume farming operations promptly. To that end, in May 2011 it created a scheme to offer financial aid to farmers affected by the disasters to help them start operating their farms again. The scheme pays 35,000 yen per 10 ares of paddy field to enable the farmers to restore their land to its previous condition to resume farming. Such restoration work might involve relatively straightforward removal of garbage and debris, repairing ridges between rice paddies and water conduits, or removing salt. This aid scheme is intended to encourage resumption of farming operations by enabling farmers to obtain income in return for undertaking farmland recovery work themselves. The financial aid is paid via agricultural recovery associations formed in each individual area, and payments are made to the individuals who make up the associations according to the role they play in recovery activities. Farmers in the city of Soma established 17 recovery associations, more than in other areas. Our project team investigated what activities were being undertaken by these agricultural recovery associations and identified the challenges they faced in pursuing their activities.

During November 12–21, 2011, the team conducted an interview-based survey, primarily targeting representatives of the agricultural recovery associations. During the survey the team started by establishing the features of the area covered by an association, the reasons behind the demarcation of that particular area, the events leading up to the association's establishment, and its main purposes. They then ascertained the extent of the damage sustained in the association's area according to three categories: farmland; irrigation facilities such as water supply and drainage channels, and ponds; and farming machinery.

Overview of Survey Results

Table 2.1 classifies the 17 agricultural recovery associations by defining three levels of damage, based on the damage sustained to farmland, irrigation facilities, and association members' farming machinery. The team used these classifications to identify the characteristics of individual associations' activities and the challenges they faced. The key features of the outcomes thus obtained were as follows (Batdelger et al. 2012).

First, the time and cost required for recovery differed entirely according to the extent of the damage. Any one-size-fits-all recovery measures that ignored the diversity in damage sustained would therefore be of little benefit. The extent of tsunami damage varied greatly from one district to another, and some types of recovery were impossible for farmers to handle on their own, making state funding

Table 2.1 Characteristics of agricultural recovery association activities according to extent of damage

Classification of the degree of damage		O = 1	△ = 2	× = 3
Damage of farmland	Flooded period	Less than 3 weeks (short term)	Less than 3 weeks (short term)	More than 3 weeks (long term)
	Existence or nonexistence of house wreckage	None	Deposited in some farmlands	Deposited in wide area and huge amounts
	Existence or nonexistence of wreckage of the windbreak forest	None	None	Deposited in huge amounts
	The amount of sediment by tsunami	Thinly partially deposited	Thickly deposited partially	Deposited in wide area and huge amounts
Damage of irrigation facilities		No problem	Self-recoverable	Own unrecoverable
Damage of agricultural machinery		None	Fewer than half of farmers lost agricultural machinery	More than half of farmers lost agricultural machinery

Note: O small damage, △ medium damage, × huge damage

essential: these included removal and replacement of topsoil covered in tsunami sediment mixed with fragments of glass or concrete, and replacement of farming machinery and equipment for farmers who had lost everything in the tsunami.

Second, the methods and time required to recover tsunami-damaged paddy fields varied greatly according to the extent of tsunami damage to the land and irrigation facilities. Included were the extent to which key farming amenities such as irrigation channels, bunds, and farm roads had been damaged; the amount of tsunami sediment and debris accumulated on the farmland; the extent to which cultivated soil had been lost, or had subsided; the length of time the fields were flooded with seawater; and the extent of damage to farming machinery and equipment.

Third, the biggest problem was how full-time farmers affected by the disasters would earn a living on a day-to-day basis when they had no source of income from a secondary occupation. The team realized there was some confusion regarding the use of the 35,000 yen per 10 areas of paddy field paid by the government for recovering of damaged farmland. They were able to confirm that in some cases the money was not necessarily used effectively to subsidize the living expenses of previously full-time farmers or to recover damaged farmland quickly.

Having identified problems such as these, the team made use of the information to provide feedback to the city authority, and as data to support the establishment of agricultural corporations.

2.3.3 Soil Fertilization Team

Since the project started on May 1, 2011, the soil fertilization team has worked to assist the recover effort by taking immediate action to establish what techniques should be used for removing salt in farmland where tsunami damage was relatively slight. The first challenge was to test the soil in farmland where sediment had accumulated because of the ingress of seawater, to use the results to promptly come up with countermeasures, and to provide information about this process to the farmers affected by the disasters. The team's tests showed that although the salt concentration was high in the tsunami sediment accumulated on the farmland, the sediment had a greater capacity to retain fertilizer than soil itself and contained large amounts of exchangeable magnesium and potassium. Moreover, although the sediment was acid sulfate soil, it contained no harmful substances such as heavy metals or arsenic. Based on the results of this soil analysis, the soil fertilization team came up with the following plan for assisting with farmland recovery.

1. Avoid removing the layer of tsunami sediment, and instead mix it into the original topsoil
2. Remove salt using rainwater and by increasing the frequency of wet tillage before planting
3. Use steelmaking converter slag to help remove salt and counteract the acid sulfate in the soil
4. Implement measures to recover farmland in the following order: (1) tsunami-damaged strawberry greenhouses; (2) paddy fields containing no debris; (3) paddy fields containing debris or those under water for a long time

2.3.3.1 Recovering the Strawberry Greenhouses

The team recovered the strawberry greenhouses as follows (Fig. 2.6)

1. Removed salt from the soil via exposure to the rain from July through October
2. Replenished organic matter and encouraged soil aggregation by planting green manure (sorgo)
3. Planted cash crops such as spinach in the fall
4. Planted cash crops in spring 2012
5. Replanted strawberry seedlings in September 2012

2.3.3.2 Method Used to Recover Tsunami-Damaged Paddy Fields

Based on the results of tsunami sediment analysis, the team decided on the following recovering process for tsunami-damaged paddy fields and provided the farmers with relevant information. This recovery plan would later gain widespread acceptance as the Tokyo University of Agriculture Method (or the Soma Method).

May 3 — August 6 — October 17

Fig. 2.6 Recovering the strawberry greenhouses (From soil fertilization team data)

1. Mix the layer of tsunami sediment into the paddy field's original topsoil
2. Improve water permeability by means of mole drainage
3. Apply 200 kg/10 ares of converter slag when the electrical conductivity of the mixed topsoil reaches approximately 0.5 mS/cm
4. Remove salt by means of wet tillage, and plant paddy rice once irrigation and drainage facilities are restored

2.3.3.3 Method Used to Recover Farmland Contaminated by Radioactivity

In farmland where the level of radioactive contamination was low, the team recommended turning over or mixing in tsunami sediment or topsoil to reduce the cesium concentration and applying natural zeolite or converter slag to improve the soil, thereby inhibiting the absorption of radionuclide by crops.

2.3.4 Forest Recovery Team

The forest recovery team is investigating ways to recovering forest areas by collecting data and analyzing the accumulation of radionuclide in forests and the penetration of cesium into trees. To that end, the team measured ambient radiation doses and collected samples for analysis, primarily targeting forests in the city of Minamisoma. The samples comprised cherry, oak, Japanese cedar, Japanese cypress, mulberry, and Japanese elk-horn cedar. The team is also experimenting with inhibiting absorption of radioactive cesium in Japanese cedar and poplar using a variety of metal ions [K^+ (potassium ions), Cs^+ (stable cesium isotope ions), and Ba^{2+} (barium ions)].

The team is currently collecting data and analyzing how long it takes for cesium to penetrate the trunks and branches of trees to determine how to decontaminate trees that contain cesium (Fig. 2.7).

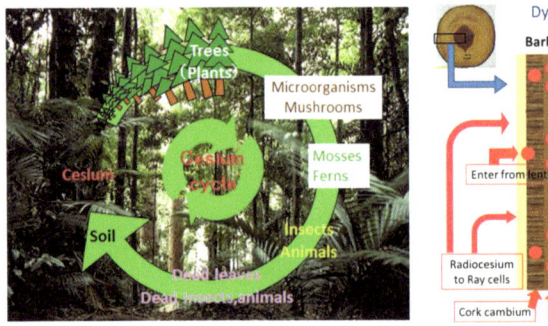
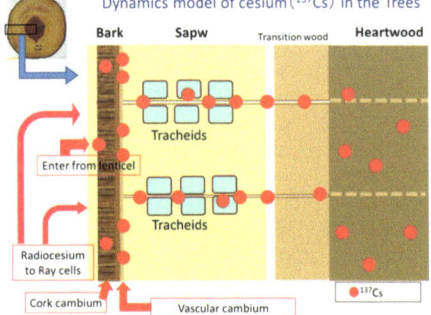

Flow of cesium-137 in the Trees

| Organization | Distribution of cesium-137 (%) | | | | |
| | Without treatment | | Fertilization of Cali | | Low temperature Short days |
	Normal	Heartwood	Normal	Heartwood	Heartwood
Leaves	34.6	32.1	+23.5	+17.5	+15.8
Tree stem					
Tip	4.7	4.5	−2.9	−3.6	−3.5
Barks	12.6	13.4	+2.9	+2.6	+2.4
Cambium	27.7	28.5	−16.1	−19.9	−25.0
Xylem	7.8	8.5	−1.8	+13.3	+6.6
Roots	12.7	13.0	−5.7	−4.9	+3.8

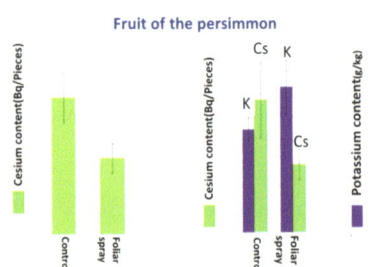

Fig. 2.7 The forest recovery team's findings (From forest recovery team data)

2.4   Success at the First Local Briefing Session Boosts Recovery Initiatives

2.4.1   The Importance of Local Briefing Sessions

From the time when we were first planning the East Japan Assistance Project we emphasized the importance of holding on-site briefing sessions about the outcomes of our initiatives. The reasons for this were as follows:

1. To win support for our research outcomes by constantly providing feedback to the farmers and organizations involved and to identify new issues
2. To win more rapid support for our research outcomes by building a relationship of greater trust with the farmers
3. To ascertain any new needs for assistance with recovery

More importantly, however, when we were working on site helping with the recovery effort, we heard quite a few farmers complaining that plenty of researchers had already been there to conduct surveys, but hardly any of them ever got back to the farmers with the results. The farmers felt that they were being treated as nothing more than research subjects. Hearing the farmers say this impressed upon us how important it was to provide feedback to the farmers on the outcomes of our research.

2.4.1.1   Anxiety and Success at the Local Briefing Session

On November 28, 2011, one half year after we had commenced our initiatives to assist the city of Soma with its recovery, we held a briefing session in Soma to report on the progress of the recovery assistance project. As project leader, I was very anxious about holding such a meeting only 6 months after we had started. I wondered how many farmers and representatives of the local organizations involved would actually turn up. I was thinking that if hardly anybody came I would have to resign as project leader, and major changes would need to be made in the project itself.

In the event, however, so many people came (around 300 in total) that the meeting room was bursting at the seams. They listened with interest to our presentations on the progress of the project, and such an enthusiastic exchange of ideas ensued that we barely had enough time to cover everything. It was an extremely productive session that left me feeling relieved that we had not been doing the wrong thing for the past 6 months, but also acutely aware of our responsibilities going forward (Fig. 2.8).

Fig. 2.8 Local farmers participated in the briefing session

2.5 Fiscal 2012 Recovery Assistance Initiatives

2.5.1 Successful Rice Harvest from Seriously Tsunami-Damaged Paddy Fields

To contribute to agricultural revival in tsunami-damaged areas, the soil fertilization team took the lead in conducting a trial to demonstrate the efficacy of using converter slag to help remove salt and counteract acid sulfate in soil as a means of effectively removing salt from tsunami-damaged paddy fields. They collaborated with farmers to plant paddy rice on 1.7 ha of paddy fields in Soma's Iwanoko district, which had sustained serious tsunami damage, and achieved a bigger-than-usual harvest of approximately 10 t. Moreover, every bag of the harvested rice passed the radioactivity tests conducted by the Fukushima Prefectural authorities, and our university's germanium semiconductor detector also detected no radioactivity. We named the rice Soma Reconstruction Rice, and marketed it actively to let the public know that the disaster zones were recovering (Fig. 2.9).

In fiscal 2013, we obtained a donation of 450 t of converter slag from Nippon Steel & Sumitomo Metal Corporation in order to use the Tokyo University of Agriculture Method to further boost the revival of paddy fields damaged by the tsunami. Together with 50 t of converter slag bought by the university, this enabled us to conduct the same trial again on a 50-ha area of paddy fields. As of the date of writing, the rice plants are growing well.

Fig. 2.9 A bag of Soma reconstruction rice (*left*) and the rice on sale (*right*)

2.5.2   Farming Assistance for Newly Established Agricultural Corporations

The Soma City authority envisaged that key responsibility for postdisaster recovery would be assumed by agricultural corporations. It asked the university to help set up these corporations, and to assist with the corporations' subsequent farming activities. The university's farming and soil fertilization teams therefore took the lead in working to support Iitoyo Farm, the city of Soma's first agricultural corporation. Support offered by the farming team included (1) providing information about setting up the corporation; (2) researching the wishes of the local farmers to determine how farming activities should be developed once the corporation was established; and (3) making observational visits to localities leading the way in terms of collaboration between agriculture, commerce, and industry, and particularly in the vertical integration of agricultural production, processing, and sales and distribution. The soil fertilization team focused on the production of soybeans, Iitoyo Farm's main agricultural activity, using a 10-ha soybean field to analyze soil and provide advice on improving its quality.

2.5.3   Developing a Monitoring System to Aid Recovery Following Radioactive Contamination

In fiscal 2012 we launched a new initiative to hasten recovery in Soma's Tamano district, where radioactive contamination was severe, and to improve the efficiency of the recovery process. The farming team took responsibility for this initiative, aiming to establish a practicable monitoring system that would ensure the safety of

Fig. 2.10 Equipment used to measure and analyze radionuclides

Fig. 2.11 Development of a system to monitor radionuclides: taking measurements (*left*), and mapping results (*right*)

all agricultural products produced in or shipped from contaminated areas. As a result, they are developing a monitoring system that enables the team to decide on appropriate decontamination measures and assess the efficacy of those measures by collecting and analyzing such key data as the ambient and soil radiation doses, the depth of the topsoil, and soil characteristics, one parcel of land at a time. They have already collected and analyzed data for the whole of Tamano, including its 646 paddy fields, non-paddy arable fields, pastures, and greenhouses, and have drawn up a decontamination plan (Figs. 2.10 and 2.11).

2.5.4 *Identifying and Counteracting Reputational Damage*

To understand the extent of damage to the reputation of agricultural produce within Fukushima Prefecture, and to find a way of overcoming that negative reputation, in December 2011 we surveyed 200 consumers at the Aizu Wakamatsu Farmers' Market. For the first year after the nuclear disaster, the national government had set a provisional limit for radioactivity in nondairy agricultural products of 500 Bq/kg or less, and in April 2012 it set a new, reduced limit of 100 Bq/kg or less. During the survey, as many as 40–50 % of respondents replied that they could not decide whether these limits were safe, and because the meanings of the terms used to describe these limits had not been explained clearly to consumers, they had no means of judging their safety. In addition, we found that they were unsure about the reliability of the actual figures used for the limits. Although reducing the limit from 500 Bq/kg to 100 Bq/kg could no doubt make consumers feel safer to a certain extent, still almost one in four of our respondents replied that they did not feel reassured by the new limit. This response indicated that it was difficult to eradicate consumers' anxiety just by reducing the limit. It was clear, therefore, that at the very least agricultural products needed to obtain an "ND" (not detectable) rating in tests if consumers' peace of mind was to be assured (Monma 2014 Fig. 2.12).

2.5.5 Investigating Methods for Inhibiting Absorption of Radionuclide

The soil fertilization team is working both to recover farmland damaged by the tsunami and to develop and trial practicable methods for decontaminating farmland contaminated with radionuclide. It is playing an important role in initiatives to develop decontamination methods through its analysis of soil fertilization. One such initiative targets the paddy fields in Minamisoma, where planting of paddy rice for consumption is prohibited; in another case the team is working with the University of Tokyo and Fukushima University to develop an efficient means of decontaminating paddy fields in the city of Date (Fig. 2.13).

2.5.6 Forest Reconstruction Initiatives

The forest reconstruction team is pursuing research to help establish measures for counteracting radioactive contamination in forests. To this end, it is investigating how radionuclides enter trees via their leaves, bark, and roots to contaminate the wood itself. Specifically, this entailed felling approximately 30 trees, primarily Japanese cedar and cypress, and cutting the wood into cross sections to analyze the accumulation of radionuclide over time according to each annual growth ring (Fig. 2.14). This plan enabled the team to ascertain that the highest levels of radioactive cesium were found in the outer bark, but that cesium had also penetrated to the wood inside.

When the team released the results of their analysis to the press on February 1, 2012, the public received the news with some alarm. Currently, the team is still analyzing in detail the routes by which cesium penetrates the trees' interior. It is also conducting empirical research into the efficacy of specific decontamination methods such as aerial spraying of potassium and washing contaminated wood with water.

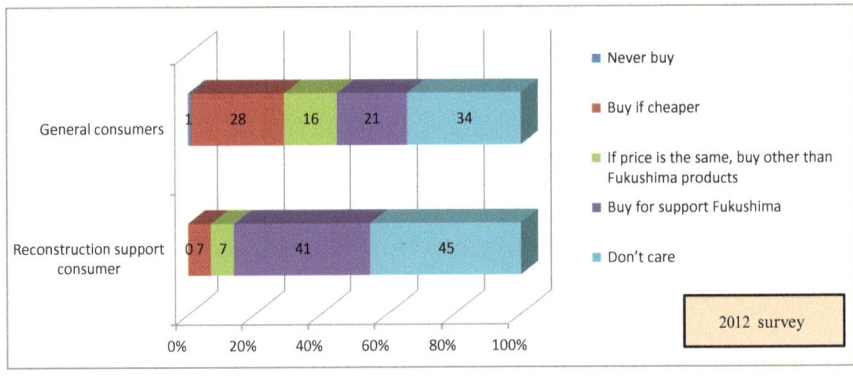

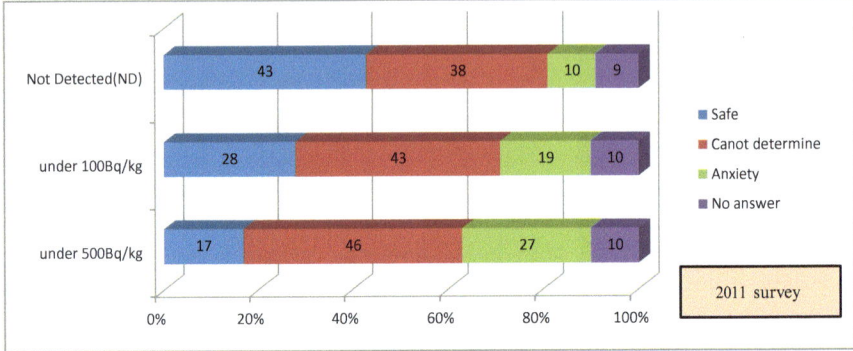

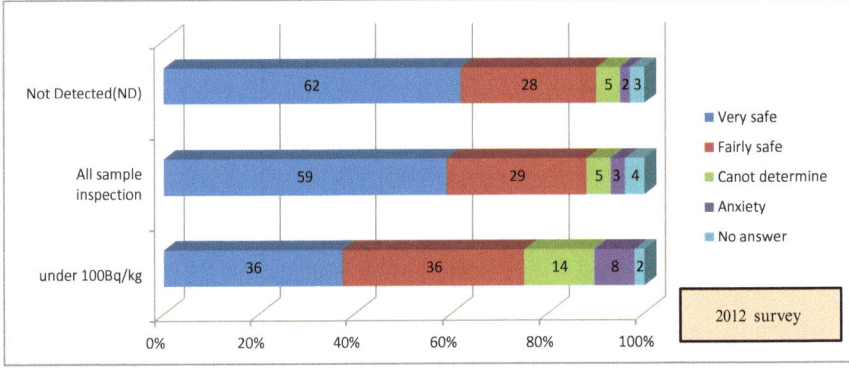

Fig. 2.12 Consumer willingness to purchase agricultural products from Fukushima and consumer sentiment toward test method and radioactive content limits

2.5.7 The Fiscal 2012 Local Briefing Sessions

The local briefing sessions on the fiscal 2012 outcomes of the East Japan Assistance Project's initiatives were held on February 23 and 24, 2013. The two sessions were held in separate venues, one in the area damaged by the tsunami, and the other in the area that had suffered radioactive contamination.

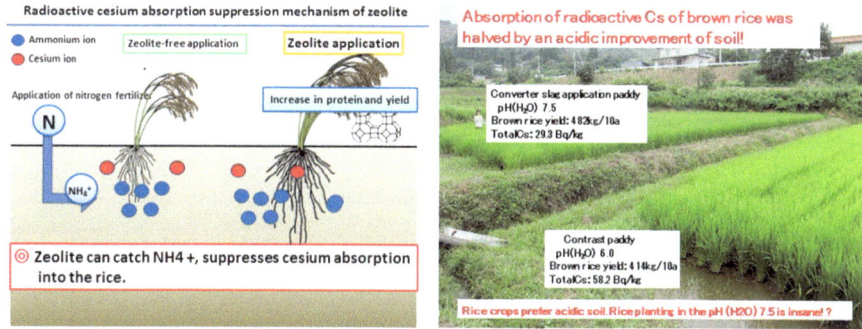

Fig. 2.13 Technology to inhibit absorption of radioactive cesium (*left*), and the technology under trial (*right*) (From soil fertilization team data provided)

Fig. 2.14 Tree growth and cesium absorption (*left*) and decontamination of a persimmon tree (*right*) (From forest reconstruction team data provided)

On February 23 approximately 200 people attended, leaving us in no doubt as to the residents' high expectations of the Tokyo University of Agriculture's efforts to help with recovery. First, our researchers addressed the initiatives to assist with reconstruction in tsunami-damaged areas, giving a presentation about the success achieved using the Tokyo University of Agriculture Method to revive seriously damaged paddy fields. Their success story provided all those present with strong motivation to plant a greater area in fiscal 2013. The residents were also interested to hear a presentation about the problematic issue of damage sustained by forests, and the associated efforts to decontaminate it. Meanwhile, another presentation reported the results of a survey on local residents' opinions and expectations with regard to the newly established agricultural corporations. It described generally high hopes that the area's agriculture could continue and thrive if new entities and individuals could be organized to take over, and indicated that rapid integration of farmland was likely to be possible. Thus the audience was left feeling hopeful with

Fig. 2.15 Briefing session held in Tamano in heavy snow

regard to new developments in Soma's agriculture. Then, moving on to the issue of reputational damage, the researchers presented the results of their surveys on consumers' opinions of, and willingness to buy, agricultural produce from Fukushima Prefecture, and on the efficacy of testing for radionuclide in terms of reducing the negative reputation. They emphasized the importance of endeavoring to make absolutely sure that no contaminated agricultural produce is shipped from Fukushima and were able to win the audience's support in this regard.

On February 24 we were scheduled to hold our first briefing in Tamano, where the biggest problem was radioactive contamination. Unfortunately, however, the temperature had dropped and it snowed heavily, so we did not expect many people to come. A low turnout seemed likely also because few people had attended most of our previous information sessions about radioactivity-related measures. But by the time the session was to start, the Tamano community center venue was packed with nearly 80 local residents (Fig. 2.15). It was clear to us, therefore, that the local people were very keen to revive agriculture in their area once decontamination was completed. Our researchers' presentations covered technology to inhibit the absorption of radioactive cesium in paddy fields, methods for decontaminating forests, radioactive contamination in the Tamano district and the results achieved using the monitoring system, as well as the situation with regard to reputational damage and measures for overcoming the negative reputation. The discussion that followed became a spirited debate between the audience and the presenters, as members of the audience asked question after question about how agricultural production could take place in future by controlling radionuclide. The snow fell rapidly as the session proceeded, reaching nearly 20 cm in depth, but everybody stayed until the end, and it turned out to be a very worthwhile briefing.

2.6 Course of Action in Fiscal 2013

The key initiatives during fiscal 2013 (April 2013 to March 2014) are as follows.

1. Work to expand our recovery activities from individual points on the map to wider geographic areas. To this end, we used the Tokyo University of Agriculture

Method to plant paddy rice in 50 ha of paddy fields that had been seriously damaged by the tsunami, and planted soybeans in the 1.7 ha of paddy field that was used for producing rice last year. We plan to ensure the safety of the rice and soybeans harvested from these fields by testing for radionuclides; all the rice confirmed as safe will be actively marketed as Soma Revival Rice.

2. Consider new ways of using paddy fields that were damaged so seriously that reconstruction is impossible and investigate models for the area's agriculture accordingly.
3. Develop and trial technology to inhibit the absorption of radionuclide in paddy fields where radioactive contamination is severe (continue joint research in the city of Date with the University of Tokyo and Fukushima University).
4. Investigate the mechanisms whereby radionuclides move through forests and trees. Use this research to develop and trial technology for eliminating radioactive cesium accumulated in persimmon trees with a view to resuming local production of partially dried persimmons. Meanwhile, amass scientific knowledge to be used in developing a variety of potentially effective methods for decontaminating the forests, and conduct on-site trials of such methods.
5. Finish developing the system for monitoring radionuclides in the city of Soma's Tamano district, and use the final version as the basis for developing and trialing a new system to control radiation levels in both soil and crops.
6. Assist with the activities of Soma's newly established agricultural corporations (by identifying ways for the corporations to collaborate with other local farmers and manage their businesses, and by helping with efforts to vertically integrate agricultural production, processing, and sales and distribution).
7. Assess how the area's agriculture might develop in future, and recommend a vision for that agriculture centered on new entities and individuals, along with a suitable governmental policy structure to provide support.

References

Batdelger N, Yamada T, Suzumura G, Sibuya Y, Lurhathaiopath P, Monma T (2012) Actual situation and evaluation of reconstruction union activities in tsunami damaged area. J Rural Econ (Special Issue 2012):192–198 (in Japanese)

Monma T (2013) Efforts of the reconstruction from Great East Japan earthquake by Tokyo University of Agriculture and Scenario of farming reconstruction. J Food Agric Environ 11:22–25 (in Japanese)

Monma T (2014) Research trends of radioactive contamination influence on agriculture and food consumption behavior. J Rural Econ 85(1):16–27 (in Japanese)

Chapter 3
Characteristics of the Agricultural and Forestry Industries in the Soma Area and Damage Sustained as a Result of the Great East Japan Earthquake

Takahiro Yamada, Puangkaew Lurhathaiopath, and Toshiyuki Monma

Abstract This chapter is organized about the characteristics of the agricultural and forestry industries in the Soma area (the cities of Soma and Minamisoma, the town of Shinchi, and the village of Iitate) and the damage sustained as a result of the Great East Japan Earthquake, by using the statistics published by the country and municipalities and related documentation. The earthquake and tsunami wrought serious damage on agriculture and agriculture- and forestry-related communal facilities in the Soma area. At the Fukushima Daiichi nuclear power station, the hydrogen explosions triggered by the earthquake and tsunami damaged reactor buildings, and large quantities of radionuclides were dispersed over a wide area. Following the accidents at the nuclear plant, two communities in the Soma area—the village of Iitate and part of the city of Minamisoma—were designated as evacuation zones. To bring an end to these consequences of the radioactive contamination as soon as possible, the city authority is implementing radiation-related measures.

Keywords Soma area • Great East Japan Earthquake

T. Yamada (✉) • T. Monma
Department of International Biobusiness Studies, Tokyo University of Agriculture,
1-1-1 Sakuragaoka, Setagaya, Tokyo 156-8502, Japan
e-mail: t4yamada@nodai.ac.jp

P. Lurhathaiopath
Faculty of Life and Environmental Sciences, Tsukuba University,
1-1-1 Tennodai, Tsukuba City, Ibaraki Prefecture 305-8577, Japan

© The Author(s) 2015
T. Monma et al. (eds.), *Agricultural and Forestry Reconstruction*
After the Great East Japan Earthquake, DOI 10.1007/978-4-431-55558-2_3

43

3.1 Characteristics of the Agricultural and Forestry Industries in the Soma Area Before the Great East Japan Earthquake

3.1.1 Characteristics of the Soma Area

The Soma area in Fukushima Prefecture comprises the cities of Soma and Minamisoma, the town of Shinchi, and the village of Iitate, located in the northeast of the prefecture. Soma, Minamisoma, and Shinchi are on the Pacific coast, and Iitate is inland in the Abukuma Highlands, an area benefiting from abundant natural assets. In terms of climate, the seasonal winds that blow from the Japan Sea are cut off by the Abukuma mountains, so, in contrast to much of Japan's northeastern Tohoku region, the Pacific coastal area in particular receives comparatively little snowfall in winter, making it a pleasant location in which to live.

Figure 3.1 shows the economies of the municipalities that make up the Soma area. Total economic output by municipality ranges from highest to lowest as follows: Minamisoma. 250,111 million yen; Soma City, 146,521 million yen; Shinchi, 41,083 million yen; Iitate, 12,781 million yen. Averaged across the municipalities of Fukushima Prefecture as a whole, primary industry accounts for 2.1 % of total economic output, secondary industry accounts for 27.9 %, and tertiary industry for 69.6 %. In Minamisoma and Shinchi the distribution is similar, but in these two municipalities tertiary industry accounts for a higher percentage of economic output than in Fukushima Prefecture as a whole, with the electricity, gas, and water utility industries accounting for a particularly high share. In the city of Soma, on the other hand, secondary industry accounts for a higher percentage (48.7 %) of economic output than tertiary industry (47.4 %), with manufacturing accounting for a particularly high percentage (42.6 %) of the total economic output. In Iitate, primary industry commands a relatively high share (14.4 %) of the total economic output, although its share is still less than that of the secondary or tertiary industries. Agriculture's share of Iitate's total output is relatively high at 13.7 %, and agriculture could therefore be described as one of the village's key industries.

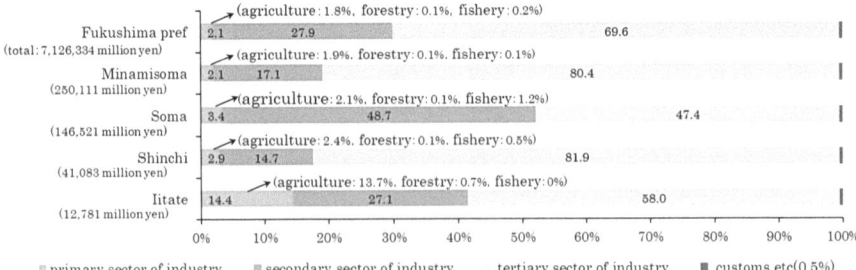

Fig. 3.1 Distribution of total economic output by municipality and economic sector in Fukushima Prefecture (fiscal 2010). (From Prefectural Accounts in Fukushima Prefecture for fiscal 2010)

3.1.2 Characteristics of the Agricultural and Forestry Industries in the Soma Area

Table 3.1 shows the area of cultivated land under management and the number of farm management entities in Fukushima Prefecture by area. The first observation from the table is that the Nakadori region accounts for the largest share (63,467 ha) of the total area of Fukushima Prefecture of cultivated land under management (121,488 ha). Nakadori is followed by the Aizu region (31,242 ha) and the Hamadori region (26,684 ha). By type, cultivated land in Fukushima Prefecture is made up of paddy fields (90,572 ha), non-paddy arable fields (25,057 ha), and land under permanent crops (5,859 ha). The area of cultivated land under management in the Soma area comprises 7,486 ha in Minamisoma, 3,123 ha in Soma city, 2,331 ha in Iitate, and 1,122 ha in Shinchi. In terms of land category, Minamisoma, Soma city, and Shinchi—coastal areas featuring extensive plains—have the prefecture's highest ratios of paddy fields cultivated, with 84.1 %, 88.0 %, and 73.3 %, respectively. Iitate has a high ratio of non-paddy arable field area under cultivation (49.5 %) because it is located on a plateau. Fukushima Prefecture as a whole has 71,091 farming entities that own cultivated land, and by district the breakdown is Nakadori 42,528 ha, Hamadori 14,577 ha, and Aizu 13,917 ha. Overall, approximately 90 % of farming entities that own cultivated land in Fukushima Prefecture own paddy fields and non-paddy arable fields.

Table 3.2 depicts the Soma area's forestland by form of ownership. Overall, Fukushima Prefecture has 971,694 ha of forest. Within the Soma area, the forest is divided among the municipalities as follows: Minamisoma 21,947 ha, Iitate 17,531 ha, Soma City 10,143 ha, and Shinchi 1,636 ha; the Soma area as a whole accounts for approximately 5.3 % of Fukushima Prefecture's forested land. By form of ownership, forest under the jurisdiction of the Japanese government's Forestry Agency (national forest) accounts for the highest share across Fukushima Prefecture as a whole with 41.6 %. In the Soma area, 58 % of Iitate's forest is under the jurisdiction of the Forestry Agency; the percentages of private forest owned by individuals and others are high in Shinchi (67.1 %), Minamisoma (42.1 %), and Soma City (41.3 %).

Next, we examine the changes in the number of hectares managed by the Soma area's farming and forestry entities between 2005 and 2010 (Table 3.3). In terms of farming entities, across Fukushima Prefecture as a whole, the number of relatively small operations managing less than 3 ha decreased during the 5 years whereas the number of entities managing 5 ha or more increased. Entities managing 20 to 30 ha and 30 to 50 ha in particular show a marked increase in number, indicating that the prefecture's farming entities are on the whole increasing the size of their operations. In the Soma area, the number of farming entities managing less than 5 ha generally decreased whereas the number of entities managing 5 ha or more increased.

Table 3.1 Area of cultivated land under management and number of farming entities in Fukushima Prefecture (fiscal 2010)

		Area of cultivated land under management	Paddy fields	Non-paddy arable fields	Fruit orchards	Management entities of cultivated land	Management entities of paddy fields	Management entities of non-paddy fields	Management entities of fruit orchards
Fukushima Prefecture	Number	121,488	90,572	25,057	5,859	71,091	67,956	61,236	9,353
	Percentage	100	74.6	20.6	4.8	100	95.6	86.1	13.2
Nakadori	Number	63,467	43,269	15,269	4,935	42,528	40,330	36,295	7,381
	Percentage	100	68.2	24.1	7.8	100	94.8	85.3	17.4
Aizu	Number	31,242	26,427	4,244	572	13,917	13,494	12,586	1,287
	Percentage	100	84.6	13.6	1.8	100	97.0	90.4	9.2
Hamadori	Number	26,684	20,828	5,505	354	14,577	14,064	12,288	678
	Percentage	100	78.1	20.6	1.3	100	96.5	84.3	4.7
Soma area	Minami soma	Number	7,486	6,293	1,139	54	3,052	2,954	2,616	106
		Percentage	100	84.1	15.2	0.7	100	96.8	85.7	3.5
	Soma	Number	3,123	2,748	312	64	1,278	1,243	1,038	77
		Percentage	100	88.0	10.0	2.0	100	97.3	81.2	6.0
	Shinchi	Number	1,122	822	275	25	530	489	492	59
		Percentage	100	73.3	24.5	2.2	100	92.3	92.8	11.1
	Iitate	Number	2,331	1,173	1,155	4	763	720	694	18
		Percentage	100	50.3	49.5	0.2	100	94.4	91.0	2.4

Source: Census of Agriculture and Forestry 2010
Note: Numerical value of Fukushima Prefecture and the total value of each region does not match because unpublished data exist

Table 3.2 Soma area's forests by municipality and form of ownership (fiscal 2010) (unit: ha, percentage)

| | | Total | National forest | | Public forest | | | | Private | | | | Center for forestry and agriculture development |
			Forestry agency	Othe	Prefecture	Public corporation	City	Property ward	Company	Temples and shrines	Joint holding	Individual	
Fukushima Prefecture	Number	971,694	404,345	4,621	10,776	15,842	42,243	24,275	25,969	4,228	140,918	286,079	12,396
	Percentage	100	41.6	0.5	1.1	1.6	4.3	2.5	2.7	0.4	14.5	29.4	1.3
Minami soma	Number	21,947	8,772	136	485	15	277	54	240	48	2,638	9,237	45
	Percentage	100	40.0	0.6	2.2	0.1	1.3	0.2	1.1	0.2	12.0	42.1	0.2
Soma	Number	10,143	2,757	–	307	–	99	14	813	9	1,907	4,186	51
	Percentage	100	27.2	–	3.0	–	1.0	0.1	8.0	0.1	18.8	41.3	0.5
Shinchi	Number	1,636	8	–	102	–	118	–	33	1	247	1,097	29
	Percentage	100	0.5	–	6.2	–	7.2	–	2.0	0.1	15.1	67.1	1.8
Iitate	Number	17,531	10,243	12	12	531	501	–	438	17	575	5,183	19
	Percentage	100	58.4	0.10	0.1	3.0	2.9	–	2.5	0.1	3.3	29.6	0.1

Source: Statistical yearbook of Fukushima Prefecture (127th)

Table 3.3 Farm and forestry management entities in the Soma area by number (no.) of hectares under management (2010) (unit: number of company (entity), percentage)

		Farm management entities	Less than 1 ha	1–1.5 ha	1.5–2 ha	2–3 ha	4–5 ha	5–10 ha	10–20 ha	20–30 ha	30–50 ha	50–100 ha	100 ha or more
Fukushima prefecture	2005	81,791	39,500	15,534	9,375	9,436	5,367	2,057	429	55	20	16	2
	2010	71,654	31,755	13,583	8,421	8,727	5,399	2,417	607	109	45	27	1
	Increase/decrease	−12	−20	−13	−10	−8	1	18	41	98	125	69	−50
Minami soma	2005	3,708	1,178	683	575	662	404	150	46	6	4	0	0
	2010	3,086	871	584	445	552	352	153	70	15	7	3	–
	Increase/decrease	−17	−26	−14	−23	−17	−13	2	52	150	75	–	–
Soma	2005	1,573	463	297	243	287	183	74	25	1	–	–	–
	2010	1,285	336	218	195	244	168	82	31	4	–	–	–
	Increase/decrease	−18	−27	−27	−20	−15	−8	11	24	300	–	–	–
Shinchi	2005	687	310	159	97	57	36	21	5	2	–	–	–
	2010	536	203	126	79	57	30	21	7	3	3	1	–
	Increase/decrease	−22	−35	−21	−19	0	−17	0	40	50	–	–	–
Iitate	2005	928	296	220	146	153	74	22	8	2	–	7	–
	2010	771	238	151	113	115	80	36	13	7	1	9	–
	Increase/decrease	−17	−20	−31	−23	−25	8	64	63	250	–	29	–

Fukushima prefecture	2005	7,189	62	2,705	2,192	1,226	366	249	152	104	12	12
	2010	4,929	73	1,714	1,472	912	271	181	112	82	10	13
	Increase/decrease	-31	18	-37	-33	-26	-26	-27	-26	-21	-17	8
Minami soma	2005	249	3	84	73	50	16	5	9	2	1	1
	2010	210	4	59	72	44	14	7	6	2	1	–
	Increase/decrease	-16	33	-30	-1	-12	-13	40	-33	0	1	–
Soma	2005	326	–	171	93	36	8	8	5	2	–	–
	2010	307	–	155	86	42	8	6	6	2	–	–
	Increase/decrease	-6	–	-9	-8	17	0	-25	20	0	–	–
Shinchi	2005	12	–	6	3	1	–	1	1	–	–	–
	2010	2	×	×	×	×	×	×	×	×	×	×
	Increase/decrease	-83	–	–	–	–	–	–	–	–	–	–
Iitate	2005	84	–	26	28	16	7	4	2	–	1	–
	2010	107	2	34	35	21	9	3	2	–	1	–
	Increase/decrease	27	–	31	25	31	29	-25	0	–	0	–

Source: Census of Agriculture and Forestry 2010
Note: '%' indicates the rate of increase or decrease from 2005 to 2010

Minamisoma, Shinchi, and Iitate show early signs that large farming entities managing 50 ha or more are starting to appear. Nonetheless, throughout Fukushima Prefecture and the Soma area, small- to medium-sized farming entities and entities managing less than 5 ha account for the majority.

Turning to forestry entities, the number in Fukushima Prefecture as a whole decreased dramatically (by 31 %) during the 5 years. Forestry entities managing less than 1,000 ha decreased in number overall, with the sole exception of small entities managing less than 3 ha, which maintained an 18 % increase. The same trend prevails in the Soma area also, revealing an apparent lack of new forestry entities to take over.

3.1.3 Changes in Farming and Forestry Entities and Output

Table 3.4 shows changes between 2005 and 2010 in the numbers of farming and forestry entities in the Soma area by type of management structure. From this table we see that the total number of farming entities decreased within Fukushima Prefecture and the Soma area, but the number of incorporated farming entities increased. Across the prefecture, cooperative corporations and joint-stock companies in particular rose in number. Within the Soma area, Minamisoma witnessed conspicuous growth in the number of farming entities becoming joint-stock companies. Despite this increase, however, the majority of farming entities in the Soma area remain individually operated or unincorporated.

Table 3.5 depicts the monetary values of the Soma area's agricultural output by product in fiscal 2008, and in fiscal 2011, the year of the earthquake. We observe from the table that in fiscal 2008 the product generating the highest output in terms of monetary value was rice, followed by vegetables. By municipality, Minamisoma (5.5 billion yen) and Shinchi (710 million yen) had the highest rice output in monetary terms. In the city of Soma, on the other hand, the output of vegetables was valued highest at 4.87 billion yen, with rice next at 2.41 billion yen. In the village of Iitate, located in a mountainous region, the outputs of rice (680 million yen) and beef cattle (640 million yen) were highest.

Data relating to agricultural output in fiscal 2011, the year of the Great East Japan Earthquake, are disclosed only for the city of Soma and the village of Iitate. In Soma, we see a dramatic drop not only in the city's overall agricultural output, but also in the output of its former staple products, rice and vegetables, as well as in other products. Similarly, in Iitate, the village's total agricultural output for fiscal 2011 was worth less than half of its fiscal 2008 value.

Table 3.4 Farm and forestry management entities in the Soma area by type of management structure (2005 and 2010) (unit: number of company (entity), percentage)

| | | Farm management entities | | | | | Forestry management entities | | | | | | | |
| | | Total | Incorporation | | Company | | Total | Incorporation | | | Forest cooperative | Other groups | Other | Administration, other |
			Corporation total	Cooperative corporation	Co. Ltd	Inc. Ltd		Corpo ration total	Company Co. Ltd	Inc. Ltd				
Fukushima prefecture	2005	81,791	525	80	36	237	7,100	248	30	72	45	54	30	89
	2010	71,654	616	109	307	4	4,853	178	67	4	41	37	9	76
Minami soma	2005	3,708	26	1	2	14	248	9	1	4	1	2	0	1
	2010	3,086	39	6	20	–	210	5	3	3	1	–	–	–
Soma	2005	1,573	6	–	2	2	324	7	3	2	1	–	1	2
	2010	1,285	9	1	4	–	306	7	3	–	1	1	1	1
Shinchi	2005	687	4	–	–	4	–	–	–	–	–	–	–	–
	2010	536	5	–	5	–	–	×	×	–	×	×	×	×
Iitate	2005	928	13	10	–	1	83	8	–	–	1	5	–	1
	2010	771	25	8	3	–	106	5	–	–	1	3	–	1

Source: Census of Agriculture and Forestry 2010

Table 3.5 Monetary values of Soma area's agricultural output by product (fiscal 2008 and 2011) (unit: 10 million yen)

		Total	Rice	Vegetable	Fruits	Other crop cultivation	Beef cattle	Dairy cattle	Hog	Poultry
Fukushima prefecture	2008	1,003	550	172	21	11	84	56	50	×
Soma	2008	988	241	487	30	3	16	28	12	156
	2011	682	137	323	22	11	9	21	16	×
Shinchi	2008	193	71	68	10	–	0	7	–	×
Iitate	2008	362	68	42	0	1	64	17	×	×
	2011	140	38	57	9	16	–	5	–	×

Source: Statistic of gross values of production for agriculture in Fukushima Prefecture (2006 and 2011)

3.2 Damage to the Soma Area's Agricultural and Forestry Industries as a Result of the Great East Japan Earthquake

3.2.1 Loss of Life and Agricultural Damage Caused by the Tsunami

Approximately 1 h after the Great East Japan Earthquake on March 11, 2011, a tsunami reached the port of the city of Soma. The tsunami, which inundated the city relentlessly, reached a maximum wave height of 9.3 m. Along the Soma area coast, the tsunami smashed through the forest of trees more than 10 m tall that had been planted on the shoreline to protect against such an eventuality. At a stroke the water swallowed up streets full of homes and shops, fishing harbors, agricultural land, roads, and railroad tracks located in low-lying areas up to 2 km inland. Vast quantities of debris, and sludge from the ocean floor, were deposited in the communities damaged by the tsunami. In addition, the earthquake caused liquefaction and subsidence of land reclaimed through infilling or drainage throughout the Soma area. It was reported that land reclaimed by drainage in the city of Soma itself subsided by some 39 cm, and this subsidence caused flood damage whenever high waves, spring tides, or heavy rain occurred subsequently (Fig. 3.2).

What is more, a great many people in the Soma area lost their lives to the Great East Japan Earthquake and ensuing tsunami, and many others were deprived of the communities in which they had lived. As of 2013, a total of 1,700 people were known to have died in the Soma area as a result of the earthquake and tsunami, and the number of homes known to have been partially damaged or completely destroyed had reached approximately 9,500 (affecting some 4,600 families).

Meanwhile, at the Fukushima Daiichi nuclear power station operated by Tokyo Electric Power Company (TEPCO), the hydrogen explosions triggered by the earthquake and tsunami damaged reactor buildings, and large quantities of radionuclides

Fig. 3.2 Paddy fields and farming machinery damaged by the tsunami

Table 3.6 Farmland in the Soma area washed away or submerged by the Tsunami (2011) (units: ha, percentage)

	Area of cultivated land under management (2010)	Estimated damage area by the Tsunami	Percentage	Estimated damage area of paddy and non-paddy arable fields	
				Paddy	Non-paddy arable fields
Fukushima Prefecture	149,900	5,923	4	5,588	335
Minamisoma	8,400	2,722	32	2,643	80
Soma	3,910	1,311	34	1,251	60
Shinchi	1,330	433	33	428	5

Source: Ministry of Agriculture, Forestry and Fisheries (2011) Estimated areas of agricultural land damaged by washing away, flooding, etc., caused by the tsunami

were dispersed over a wide area. Following the accidents at the nuclear plant, two communities in the Soma area—the village of Iitate and part of the city of Minamisoma—were designated as evacuation zones. In addition, Iitate's local government offices were forced to relocate to the city of Fukushima. At the same time, many residents who had lost their homes as a result of the earthquake and tsunami left the area in search of somewhere to live long term, such as temporary accommodation inside or outside Fukushima Prefecture, relatives' homes, or rented apartments.

The earthquake and tsunami also wrought serious damage on agriculture in the Soma area. Table 3.6 shows the extent to which farmland in the Soma area was washed away or submerged by the tsunami. We see from the table that the estimated total area of farmland damaged in the prefecture as a whole was as much as 5,923 ha, of which 5,588 ha was paddy fields and 335 ha was non-paddy arable fields.

Within the Soma area, approximately 4,400 ha of farmland was damaged in the three tsunami-ravaged municipalities of Minamisoma, Soma City, and Shinchi.

In other words, approximately 80 % of all Fukushima Prefecture's farmland damaged by the tsunami was concentrated in the Soma area. Moreover, approximately 97 %, or 4,322 ha, of the tsunami-damaged farmland in the Soma area comprised paddy fields.

For farmers, machinery and equipment are indispensable assets for operating their farms. No statistics on damage to farming machinery and equipment are available, but it can be surmised that most of the farmers who lost their homes in the tsunami would have lost their machinery and equipment at the same time. Their immediate need was to reestablish their livelihood, yet they lacked the financial means to buy new machinery and equipment, and this contributed to dwindling motivation to resume farming.

Moreover, farmers faced another serious problem in the form of damage to agriculture- and forestry-related communal facilities.

Table 3.7 shows data on damage across Fukushima Prefecture as a whole, where the cost of damage to farmland and agricultural facilities was estimated to have reached approximately 27.3 billion yen. The damage caused by the earthquake and

Table 3.7 Damage to agriculture- and forestry-related communal facilities in Fukushima Prefecture (2011)

Agriculture			Forestry		
Classification	Number	Amount of damage (million yen)	Classification	Number	Amount of damage (million yen)
Damage of agriculture	300	2,110	Damage of forestry	735	2,362
Crops	101	805	Forest	11	265
Facility in farm	199	1,305	Forest product	39	146
Damage of farmland, watercourse, other	4,358	230,258	Facility	52	1,162
Farmland	1,283	93,507	Road	633	789
Watercourse	1,133	27,491	Damage of forestland	113	14,253
Road	894	2,966			
Reservoir	745	23,611	Forest land	103	10,681
Weir	59	3,125	Facility	10	3,572
Water pump	113	28,624			
Bridge	4	84			
Embankment	2	3,000			
Irrigation and drainage channel	105	22,431			
Coastal conservation facility	20	25,419			

Source: Fukushima Prefecture homepage: Damage of public facilities such as agriculture, forestry and fisheries (2013)
Note: It does not include damage caused by the Fukushima Daiichi nuclear disaster

tsunami included flooded farmland, collapsed reservoirs, ruptured irrigation and drainage channels, and destruction of drainage pump stations. Of the total estimated cost, more than 80 % related to damage in the Hamadori region. Even if the farmers' land, machinery, and equipment had escaped damage as a result of the tsunami, therefore, they would be unable to resume farming promptly unless irrigation facilities were restored.

In addition to damage to agricultural facilities, the damage to forestry-related communal facilities was also severe. In the Nakadori region in particular, 248 forest roads sustained damage including collapsed embankments and shoulders in 633 places overall. Other types of damage included mushroom bed logs falling from shelves at shiitake-growing facilities that operate within the forestry sector. The data shown do not include damage resulting from the nuclear accident, and it is anticipated that the cost of damage including that caused by the negative reputation will increase.

3.2.2   Contamination from Radionuclides in the City of Soma and Its Effects on Agriculture and Forestry

After the accident at the Fukushima Daiichi nuclear power station, the southerly wind blowing onto the Pacific coast in the southeast of the Tohoku region carried the radionuclides emitted from the power station toward the northwest, dispersing them all over Fukushima Prefecture. Consequently, high ambient radiation doses were measured in many of the Fukushima Prefecture municipalities. Most notably, ambient radiation doses exceeding 10 μSv/h were recorded within a 20-km radius of the power station in the towns of Naraha, Tomioka, Okuma, Futaba, and Namie.

In the city of Soma, the local authority measured ambient radiation doses in front of the city hall's branch office immediately after the disaster, recording a maximum dose of 1.73 μSv/h. Although the doses gradually decreased thereafter, as of September 2013 there are still some scattered locations within Soma where the annual cumulative radiation dose exceeds 1 mSv. The local authority divided the city into a grid of 1-km squares to measure ambient radiation levels, and the results show that in June 2011 the average ambient radiation dose for the city of Soma as a whole was 0.74 μSv/h. Of the eight districts comprising the city of Soma, the highest dose was 1.88 μSv/h, recorded in the Tamano district adjacent to the village of Iitate. In addition, Yamakami district, located between Tamano district and Soma's city center, recorded a relatively high dose of 1.03 μSv/h. In the tsunami-damaged districts of Iitoyo district, Nittaki district, and Isobe district on Soma's coast, the ambient radiation dose was around 0.40 μSv/h, and the damage caused by radioactive contamination was relatively insignificant. Following the nuclear disaster the ambient radiation doses in all the districts of the city of Soma declined over time, decreasing to about half their original levels, but even now, two and a half years later, the dose in the district of Tamano district is still high, at 0.93 μSv/h (Table 3.8).

Table 3.8 Results of grid survey of ambient radiation doses in the city of Soma (unit: μSv/h)

	2011		2012		2013	
	Soil	Asphalt pavement	Soil	Asphalt pavement	Soil	Asphalt pavement
Soma City	0.74	0.60	0.53	0.36	0.36	0.24
Nakamura district	0.49	0.36	0.32	0.23	0.23	0.16
Ono district	0.48	0.38	0.37	0.27	0.25	0.17
Iitoyo district	0.39	0.34	0.22	0.18	0.18	0.12
Hachiman district	0.72	0.57	0.51	0.34	0.36	0.22
Yamakami district	1.03	0.74	0.64	0.41	0.47	0.29
Nittaki district	0.55	0.46	0.37	0.27	0.29	0.20
Isobe district	0.38	0.28	0.27	0.19	0.20	0.14
Tamano district	1.88	1.70	1.56	1.00	0.93	0.60

Source: Soma City Homepage (Information about Radioactive)
Note: (1) The survey of year 2011 was conducted on 18th June, of 2012 was on 26th April until 7th May, of 2013 was on 26th April until 10th May
(2) Soma City divided the area into square of 1×1 km, and measured the air dose rates from 173 spots at 1-m height for 10 s long for 5 times

As a result of the accident at the Fukushima Daiichi nuclear power station, many Soma residents who had lost their homes also had to live with anxiety about the radiation. Meanwhile, others left the city voluntarily because they were worried about the possible effects of radiation on their children's health. As a result of the radionuclides dispersed into the atmosphere, steps were taken to halt shipments of Soma city's agricultural produce, including vegetables and beef cattle, and planting of rice crops for consumption was prohibited. Meanwhile, highly contaminated water was released into the sea, so that fishermen who had managed to overcome the tsunami damage enough to start rebuilding their livelihoods were forced to refrain from actually fishing.

To bring an end to these consequences of the radioactive contamination as soon as possible, the city authority is implementing the following radiation-related measures.

(a) Holding information sessions in the city to enable the citizens to acquire accurate information about radioactivity.
(b) Measuring radioactivity continuously across all Soma city's districts based on 1-km grid units to gauge radiation doses and identify "hotspots," and publishing the results promptly on the city's website and in its newsletter.
(c) Continuously measuring radiation levels in 50 locations at each school in Soma to identify "mini-hotspots" and gain a detailed picture of radiation doses in the schools, and offering lectures about radiation to staff to enable them to communicate accurate information to the pupils.
(d) Taking especially detailed grid-based measurements in the district of Tamano, where radiation doses were high, and decontaminating wherever necessary. The city authority is also holding seminars and training sessions on correct decontamination methods and offering health consultations and priority admis-

sion to temporary accommodation to alleviate people's concerns about the health risks of living in Tamano district.

(e) Decontaminating with the help of local residents, using the radioactive substance decontamination manual in line with the city of Soma's decontamination plan released on December 28, 2011.

(f) Setting up a decontamination project team to implement the decontamination plan within the city, revise the manual, and verify the benefits of decontamination. In addition, the city authority is putting in place a structure to devise and implement specific measures to prioritize protecting children from radiation exposure and maintaining their health. To this end it is setting up a special committee on health measures to discuss steps to safeguard citizens' health, and particularly the health of children.

(g) Taking a variety of measures in conjunction with related agencies to dispel the negative reputation affecting agriculture, forestry, and fisheries produce, and associated processed products, as well as the industrial manufacturing and tourism industries. The city authority is also investigating the decreases in income that operators in the relevant industries have suffered as a result of negative reputation, and is sending a claim for the necessary amount of compensation to TEPCO.

(h) The decontamination process produced waste in the form of earth and sand containing radionuclides that had to be stored where it would not endanger local citizens until it could be transported to the national government's interim storage facilities. The city authority therefore set up a temporary storage site at the industrial waste treatment plant within the city, and the earth and sand were stored there. The temporary storage site is monitored with careful attention to safety to prevent dispersal, outflow, or underground seepage of radionuclide.

(i) Giving children up to the age of 15 and pregnant women "glass badges" to measure external radiation exposure for a 3-month period to safeguard the citizens' health, and particularly the health of children.

(j) Testing ingredients before lunches are prepared at schools in the city that provide their own school lunches.

(k) Installing machines in the city hall and district community centers to measure concentration of radionuclide, thereby alleviating citizen's health worries by helping them to decide whether it is safe to eat foodstuffs including vegetables and other agricultural and fish products cultivated at home. In addition, the city authority is taking measures to improve safety and peace of mind still further by deploying whole-body counters in the city's medical institutions to monitor internal radiation exposure properly and continuously, enabling citizens to manage their health.

As described, the Soma City authority is conducting an independent grid-based survey of ambient radiation doses, as well as real-time dosimeter monitoring, and surveys of ambient radiation at communal facilities within the city. In addition, surveys are currently being conducted at 68 ambient radiation monitoring posts established across Soma on April 1, 2012, by the Ministry of Education, Culture, Sports, Science and Technology. The measurement of ambient radiation levels at communal facilities takes place in six locations within the three districts of Ono district, Nakamura district, and Nittaki district. Doses at the temporary accommodation sites in the city of Soma are low, at around 0.1 µSv/h. At elementary and junior high schools the ambient radiation dose varies from one district to another, but the doses at the Tamano elementary and junior high schools before they were decontaminated were high at 2.32 µSv/h and 2.41 µSv/h, respectively. As children are most susceptible to the effects of radionuclides, Soma prioritizes their safety. To this end, the city authority decontaminated not only kindergartens and schools, but also other communal facilities frequently used by children, as well as the roads around them. As a result, the ambient radiation doses at most decontaminated elementary and junior high schools were reduced to around 0.1 µSv/h. In many sports facilities, moreover, the doses are 0.5 µSv/h or lower.

3.2.3   Effects on Agriculture, Forestry, and Fisheries Products

Immediately after the nuclear disaster, Fukushima's prefectural authority was carefully monitoring the prefecture's produce for radionuclides and disclosing the relevant information. Monitoring is still continuing today, following upgrades to the devices and system used during fiscal 2012. This monitoring makes it possible to prohibit shipment of products that exceed the limits set for radionuclides, thereby preventing the products' distribution to the market. Table 3.9 shows radiation

Table 3.9 Results of tests to monitor radionuclide in Soma City's agriculture, forestry, and fisheries products (unit: percentage)

	2011			2012			2013		
	Above limit	Under limit	No detected	Above limit	Under limit	No detected	Above limit	Under limit	No detected
Vegetable	3	17	80	1	2	97	–	4	96
Fruit	19	28	53	–	29	71	–	44	56
Mt. vegetable and mushroom	58	24	18	19	58	23	7	86	7
Meat and chicken	2	27	71	–	5	95	–	3	97
Raw milk	25	50	25	–	–	–	–	–	–
Fish	25	48	27	6	45	49	1	28	71
Wild boar	100	–	–	100	–	–	100	–	–
Pheasant	100	–	–	100	–	–	–	–	–
Wild duck	–	100	–	–	100	–	–	100	–
Fertilizer (cow dung)	62	38	–	19	61	19	22	44	33

Source: Ministry of Health, Labour and Welfare (Survey of radionuclides in food)
Note: The above numbers are the numbers of agricultural products which exceeded the new limit, applied after 1st April 2012. The limit of vegetable, fruit, Mt. vegetable and mushroom, meat and chicken, fish, wild boar, pheasant, wild duck is 100 Bq/kg, raw milk is 50 Bq/kg, and fertilizer made from cow dung is 400 Bq/kg

monitoring results for the city of Soma's agricultural, forestry, and fisheries products from the time immediately after the nuclear disaster to the present. Immediately after the disaster a wide variety of products were found to contain radionuclide exceeding the new limit for non-dairy products of 100 Bq/kg. In particular, many fish and shellfish, edible wild plants and mushrooms, and fruits, as well as cattle manure compost, were found to exceed the limit.

Bibliography

Agriculture, Forestry and Fisheries Department of Fukushima Prefecture (2013) Record of the Great East Japan Earthquake of Agriculture, Forestry and Fisheries in Fukushima prefecture (1st edn). https://www.pref.fukushima.lg.jp/sec/36005b/norinkikaku19.html

Ministry of Agriculture, Forestry and Fisheries, Basic statistical data related to agriculture, forestry and fisheries and the Great East Japan Earthquake- mainly Iwate, Miyagi, Fukushima Prefecture. http://www.maff.go.jp/j/tokei/joho/zusetu/zusetu.html

Ministry of Agriculture, Forestry and Fisheries, About the Great East Japan Earthquake-Correspond to the damage. http://www.maff.go.jp/j/kanbo/joho/saigai/higai_taiou/index.html. Soma city (2012), Soma city Revitalization Plan ver. 1.2 (in Japanese)

Monma T (2011) Sludge and debris were deposited in paddy: Challenge of research and restoration in Soma city. Shin-jitsugaku. Journal Tokyo University of Agriculture (in Japanese)

The Tohoku Regional Agricultural Administration Office (2013) Statistical yearbook of agriculture, forestry, and fisheries in Fukushima prefecture (59th) (from 2011 to 2012)

Part II
Reconstruction from Tsunami Damage

Chapter 4
Reconstruction Support for the Farmland Struck by Tsunami

Itsuo Goto and Kaisei Inagaki

Abstract Reconstruction support of the disaster-stricken areas in the city of Soma, Fukushima Prefecture, was started in May 2011. Sediment transported from the sea by the tsunami was from 5 to 10 cm thick on surface of the paddy fields. The tsunami sediment was mixed with the original soil of the paddy fields, and mole drains were formed to improve water drainage toward the lower layer. Subsequently, rainwater alone was used for salt removal. However, the pyrite in the soil had gradually oxidized until the pH dropped to 3.8, so converter slag was applied to neutralize the sulfuric acid. In May 2012, rice was transplanted into paddies from which the salt had been removed. On September, 10.7 t brown rice was harvested. The yield of brown rice harvested per hectare was 6.3 t, about 20 % higher than the amount before the disasters.

Keywords Tsunami • Tsunami sediments • Salt removal • Converter slag • Soma City

4.1 Conditions Observed in Soma's Post-tsunami Farmland and Basic Aims of Reconstruction Initiatives

4.1.1 Reconstruction Support Starting May 2011

The authors entered the disaster-stricken areas in the city of Soma, Fukushima Prefecture, on May 1, 2011, as participants in the Tokyo University of Agriculture East Japan Assistance Project. In Soma, the support base for the project (Fig. 4.1), we witnessed the devastation of the landscape caused by the giant tsunami, and it was beyond our imagination. What struck us particularly was the large number of pine trees that had previously grown on sandbanks in Matsukawaura Lagoon and were now piled up as debris in the paddies extending to its west (Fig. 4.2). Littered

I. Goto (✉) • K. Inagaki
Department of Applied Biology and Chemistry, Tokyo University of Agriculture,
1-1-1 Sakuragaoka, Setagaya-ku, Tokyo 156-8502, Japan
e-mail: igoto@nodai.ac.jp

© The Author(s) 2015
T. Monma et al. (eds.), *Agricultural and Forestry Reconstruction
After the Great East Japan Earthquake*, DOI 10.1007/978-4-431-55558-2_4

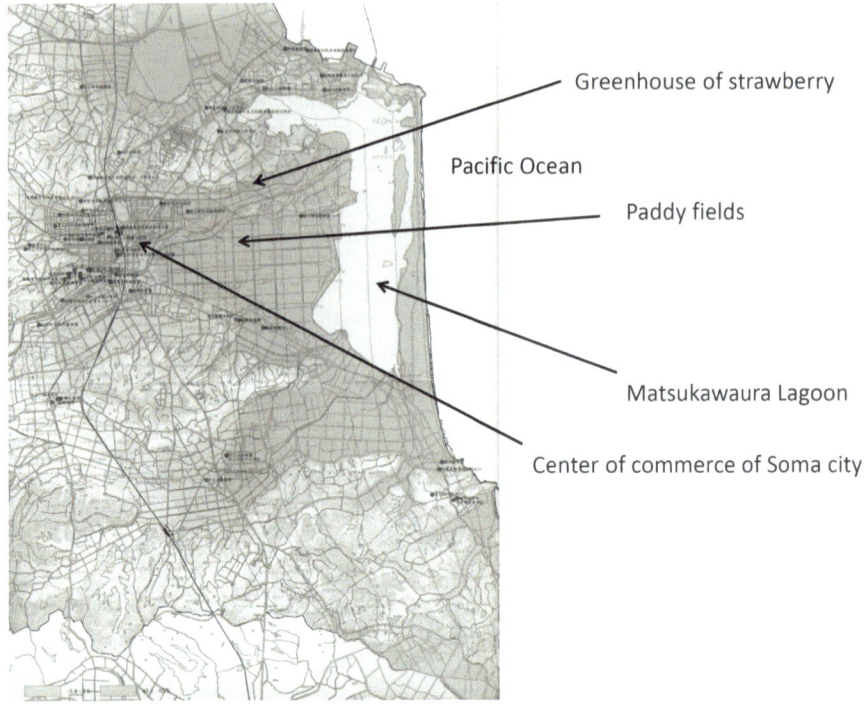

Greenhouse of strawberry

Pacific Ocean

Paddy fields

Matsukawaura Lagoon

Center of commerce of Soma city

Fig. 4.1 Map of Soma City affected by the tsunami

Fig. 4.2 Paddy fields struck by tsunami adjacent to Matsukawaura Lagoon

among these trees were cars, tractors, and washed-up fishing boats. The local farmers were devastated, bemoaning the fact that it would take years before the paddy fields could be restored to their original state. Under the huge amount of debris was a thick layer of accumulated tsunami sediment containing vast quantities of salt transported from the sea by the massive tsunami.

4.1.2 Classification of Post-tsunami Farmlands Based on the First Field Survey

Based on a field survey undertaken in May, we classified the farmlands into the following three types.

1. Paddy fields adjacent to the coast that were still flooded because of land subsidence and that contained a large amount of debris
2. Paddy fields that had been affected by the tsunami, but had started to dry up by the survey in May, were located several kilometers from the coastline and contained little debris
3. Sites such as greenhouses and upland fields that were affected by the tsunami but contained no debris

4.1.3 Properties of Tsunami Sediment Accumulated on the Surface of Tsunami-Hit Farmlands

(a) Soil chemical properties of tsunami sediment

The soil chemical properties of the tsunami sediment collected during the field survey from May 1 to 3, 2011, are shown in Table 4.1.

The sodium chloride content estimated from the exchangeable sodium content reached 2.9–5.7 %, with a high electrical conductivity (EC) of 12–24 mS/cm serving as an indication of the salt concentration. The sodium chloride content in seawater is approximately 3 %, but tsunami sediment with almost twice this amount of salt was observed (Table 4.1). However, high levels of salt content were limited to the clay layer on the surface of the tsunami sediment, with the salt content in the sand layer underneath showing a sharp progressive decline. The cation-exchange capacity (CEC) in the clay layer of the tsunami sediment was about 30 mEq/100 g, which tended to be higher than in the soil of either paddy fields or greenhouses. In addition, the soils contained a large amount of exchangeable magnesium and potassium.

Because the tsunami sediment contains substances transported from the seabed, we were concerned about the possibility of acidification similar to that found in soil on reclaimed land from the sea. We therefore measured the total sulfur content and

Table 4.1 Chemical properties of paddy soils struck by the tsunami in Soma City

Sample	Depth	pH (H$_2$O)	EC	Exchangeable bases (mg/100 g)				CEC	Degree of base Saturation (%)	Available	Available B
	(cm)		mS/cm	CaO	MgO	K$_2$O	Na$_2$O	mEq/100 g		P$_2$O$_5$ mg/100 g	mg/kg
Tsunami sediment	10	6.6	10.3	393	370	183	1,540	27.6	311	10.6	13.4
Plow layer	20	5.3	2.47	407	131	43.7	287	22.4	139	8.6	1.45
Plow sole layer	30	5.9	0.89	477	124	35.4	86.6	23.5	114	6.9	0.82
Sublayer	40	6.2	0.26	479	143	35.2	27.3	24.8	104	4.1	0.59
Sublayer	50	6.6	0.25	467	150	32.4	27.4	23.9	107	4.8	0.54

Table 4.2 Total sulfur content and pH of tsunami sediments

Sampling site	Total S %	pH (H_2O)	pH (H_2O_2)
Paddy field I	1.27	6.6	2.3
Paddy field II	1.19	6.5	2.5
Paddy field III	0.34	5.9	2.6
Paddy field IV	1.11	7.3	2.2
Upland field	1.14	7.5	2.3

Table 4.3 Concentration of trace elements of tsunami sediments and farmland soils

Classification of land	Sampling site	Sample	Depth (cm)	Cd	As	Zn	Cu	Ni	Cr
Paddy field	Kashiwazaki I	Tsunami sediment	10	0.65	8.62	162	33.2	20.6	60.0
Upland field	Kashiwazaki II	Tsunami sediment	3	0.48	8.19	124	24.6	19.0	50.3
Paddy field	Kabaniwa	Tsunami sediment	5	0.32	6.81	46.1	10.7	32.2	26.7
Greenhouse	Wada	Tsunami sediment	0.7	0.39	4.30	110	19.9	21.0	71.5
		Ridge	20	0.35	4.81	96.7	18.5	21.5	63.9
		Ridge	30	0.32	4.40	92.8	19.8	22.6	55.9
		Ridge	45	0.39	5.67	181	38.6	49.4	72.3
		Ridge	50	0.35	6.61	109	26.1	30.3	108
Greenhouse	Wada	Tsunami sediment	10	0.46	8.82	130	37.1	27.8	52.8
		Furrow	30	0.23	3.38	133	32.5	45.2	61.9
Paddy field	Wada	Tsunami sediment	5	0.43	8.97	113	32.2	22.2	47.6
		Plow layer	20	0.34	3.99	107	38.5	35.8	58.3
		Plow sole layer	30	0.28	4.31	93.7	36.1	44.1	59.1

pH (H_2O_2) of the soil collected. As shown in Table 4.2, the total sulfur contents of the tsunami sediments were about 1 % and pH (H_2O_2) was 2.2–2.6. As soil with a pH (H_2O_2) of 3 or less is defined as an acid sulfate soil, the sediment transported by the tsunami was assessed to be a potential source of acid sulfate soil.

(b) Harmful elements contained in the tsunami sediment

We analyzed the levels of harmful elements such as cadmium and arsenic in the tsunami sediment and the soil layers underneath collected from the survey sites already mentioned (Table 4.3).

The cadmium content in the tsunami sediment ranged from 0.32 to 0.65 mg/kg, with an average value of 0.39 mg/kg. The content in the soil layers underneath ranged from 0.28 to 0.39 mg/kg, with an average value of 0.29 mg/kg. Although the cadmium level in the sediment was higher than that of the underlying soil, it was equivalent to the median value of 0.39 mg/kg in farmland soil across the country.

The amount of arsenic in the tsunami sediment and soil was less than 10 mg/kg for both, which was the same as or lower than the background value in the soil. The amounts of zinc, copper, lead, nickel, and chrome in the tsunami sediment were almost the same as those in the soil.

4.2   Reconstruction Support Policy for Post-tsunami Farmlands

4.2.1   Mix Soil Layers Without Removing Tsunami Sediment

The tsunami sediment contained as much as 10–20 mS/cm of salt and a large amount of water-soluble boron (B) at 20 mg/kg, both of which substantially hinder the growth of vegetation. Meanwhile, the cation-exchange capacity (CEC) was higher than that of the plow layer, with large amounts of exchangeable magnesium and potassium serving as soil nutrients. Harmful elements such as cadmium and arsenic were a particular concern, but their levels were not found to be higher than those of the plow layer.

We knew that the salt and boron in the soil could be leached out with water, while soil acidification caused by the oxidation of pyrite could be corrected by applying lime material. We therefore decided to remove salt by mixing the sediment with the original soil and without removing the soil introduced by the tsunami.

4.2.2   Use Converter Slag as Lime Material to Remove Salt

The chloride ions in the salt are anions that exist as water-soluble ions not adsorbed by the soil colloids, but the sodium ions exist as water-soluble sodium in the form of counterions to chloride and exchangeable sodium adsorbed by the soil colloids. Therefore, when lime material is applied and mixed thoroughly with the soil, a cation-exchange reaction occurs between the calcium ions in the slag and the exchangeable sodium, allowing the exchangeable sodium to be converted to water-soluble sodium. For this reason, converter slag generated as a by-product during the steel manufacturing process in steelworks is used as lime material. The raw materials of converter slag such as iron ore, coal, and limestone contain no harmful components; the main component is calcium silicate, whereas the secondary components include micronutrients such as iron, manganese, and boron, in addition to free lime (quicklime) and magnesium. As a result, even if the pH (H_2O) is raised to about 7.5 or higher by using converter slag in the soil, it is unlikely to cause a deficiency of micronutrients in the crops. In addition, converter slag acts more slowly than calcite or dolomite in mitigating soil acidification, and can therefore also be used to counter soil acidification caused by the oxidation of pyrite in the tsunami sediment.

4.3   Reconstruction Support for Restart of Strawberry Cultivation

4.3.1   Just in Time

In the disaster zone affected by the tsunami, local farmers naturally assumed that they would have to remove the tsunami sediment, known as *hedoro* or "sludge," that had accumulated on the surface of the soil in strawberry greenhouses (Fig. 4.3). In fact, sludge removal work was scheduled to take place a few days after we first visited a strawberry farm that we were assisting. However, we managed to persuade the farmer that there was no need to remove the tsunami sediment. Instead, the first thing we did was to expose the soil to the rain by ripping off the plastic rooftops of the greenhouses and the mulching on ridges, which had been untouched since March 11. After we had finished analyzing the soil to confirm that there were no problems associated with harmful elements, the farmers themselves mixed the tsunami sediment with the original soil in June (Fig. 4.4).

Fig. 4.3 Tsunami sediment deposited on the topsoil of a strawberry greenhouse

Fig. 4.4 Mixing tsunami sediment with original soil

4.3.2 There Is No Better Salt-Remover than Rain

Instead of removing the tsunami sediment at the strawberry farm, we mixed it with the original soil on June 16. At that time, 10 t/ha of converter slag was applied as a lime material to accelerate salt removal (Fig. 4.5). The electric conductivity (EC) in the first 10 cm of the surface layer decreased from 0.64 mS/cm in July to 0.35 mS/cm in August, but the EC in the next 40-cm layer from 10 to 50 cm decreased only to the 1.5 mS/cm level. In addition, the amount of exchangeable sodium remained at 139 mg/100 g, with a high sodium percentage of 16.6 %. We therefore decided that strawberry planting should be delayed until the following year, with sorgo (a kind of green manure) planted in the meantime as a means of preparing the soil for the replanting of strawberries in September 2012. The sorgo planted on September 6, 2011 grew to almost the height of a human, reaching 1.5 m after a month (Fig. 4.6), and the yield of sorgo reached approximately 40 t/ha. After crushing with a hammer knife mower, the sorgo was plowed into the soil with a rotary plow.

As shown in Fig. 4.7, the monthly rainfall in Soma is generally highest from June to October, and 844 mm of rain fell during this period in 2011. As a consequence, the amount of salt removed using rainwater alone was greater than expected. There is no better salt remover than rain.

Fig. 4.5 Application of converter slag after mixing tsunami sediment with original soil

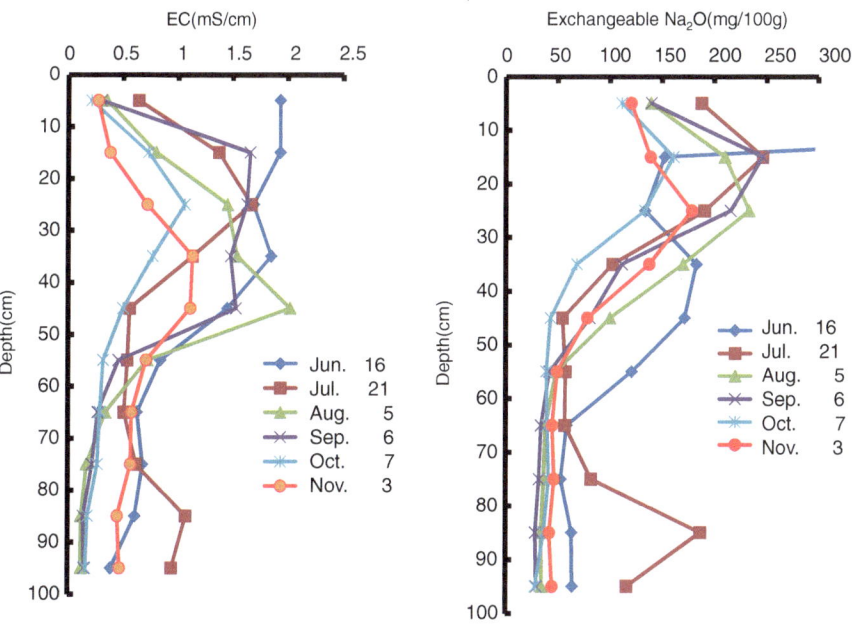

Fig. 4.6 Time-dependent changes of electric conductivity (EC) and exchangeable sodium of strawberry greenhouse soils

Fig. 4.7 Excellent growth of sorgo by 1 month after seeding

4.3.3 A Discriminating Approach to Restart of Strawberry Cultivation by Soil

Although we managed to sufficiently remove salt and adjust pH in the plow layer by September 2011, a high concentration of salt still remained in the lower layer, at a depth of about 40 cm, as shown in Fig. 4.8. If the roofs of the greenhouses were replaced and strawberries planted at this stage, it was possible that the salt in the lower layer might rise into the plow layer. As strawberries are highly susceptible to the effects of salt, we pushed the strawberry planting back by a year to fall 2012. In the meantime, we planted cash crops such as spinach, turnips, and sugar snap peas instead (Fig. 4.9).

By September 2012, the salt concentration in the lower layer had also dropped, allowing strawberry seedlings to be replanted. These subsequently grew well and harvesting started in January 2013, making these the first soil-cultivated strawberries to be harvested in this strawberry-producing area, which had been severely damaged by the tsunami (Fig. 4.10).

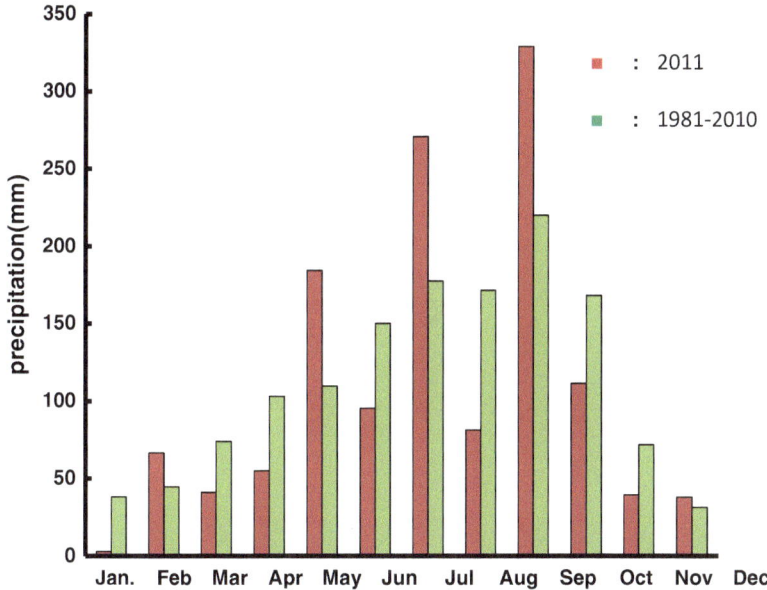

Fig. 4.8 Annual precipitation of average year and of 2011 in Soma City

Fig. 4.9 Restart of vegetable cultivation in a strawberry greenhouse (January 2011)

Fig. 4.10 Harvesting of strawberries was started from January 2013

4.4   Rejuvenating Green Manure: Also Used in Hachinohe, Aomori Prefecture

After the disasters, we were contacted by Koutaro Kimura, whom we had met at an agricultural seminar in the city of Hachinohe, Aomori Prefecture. He told us that his strawberry farm had been damaged by the tsunami (Fig. 4.11). As the Kimura farm was near the coast, the tsunami sediment was mainly sand in a layer 30 cm thick. We therefore advised him to mix this tsunami sand with the original soil using a power shovel and to apply converter slag. The basic salt removal method was the same as that employed in the strawberry farm in Soma. After salt removal by the rain, Aoba millet was planted as green manure instead of sorgo, as millet is more resistant to the *yamase*, or cold Pacific winds descending from the mountains into Hachinohe, which is situated further north than Soma. The millet planted on June 20 grew well and reached a height of 80 cm by July 20 (Fig. 4.12).

However, as shown in Fig. 4.13, growth was observed to be lacking in certain sections of the millet. We analyzed soil samples taken from these spots, and found residual salt in them; this residue resulted from inconsistencies in mixing the layers of soil using a power shovel, in contrast to Soma where the soil was mixed with a

Fig. 4.11 Strawberry greenhouse of Kimura Farm struck by the tsunami in Hachinohe City, Aomori Prefecture

Fig. 4.12 Millet thickly covered the whole area of Kimura Farm by 3 months after being struck by the tsunami (July 2011)

Fig. 4.13 A part that showed poor growth of millet was observed in Kimura Farm (*right* photograph)

rotary plow. Consequently, this discovery demonstrates the efficacy of planting green manure to confirm the removal of salt. To a farmer, such variations in growth are easier to understand than chemical analysis of the soil, and by concentrating on mixing the soil in areas that do not grow well, more uniform salt removal can be achieved. Thus, in salt-damaged farmlands, green manure serves a secondary purpose of confirming salt removal.

4.5 Taking on the Challenge of Planting "Soma Revival Rice"

4.5.1 Weeds in Salt-Damaged Paddies Convince the Farmers

Figure 4.2 shows the scene we came upon when we first visited Soma on May 1, 2011. The devastation was beyond our imagination and we thought it would probably take years to restore the land to its original state. Nonetheless, all the debris in the paddy fields had in fact been removed by September, and weeds such as barnyard grass could be seen growing here and there, even though much of the paddy surface had dried up like a tortoise shell and been reduced to a barren desert (Fig. 4.14). The tortoise shell effect was caused by sediment transported from the sea by the tsunami that had dried and cracked in a layer about 10 cm thick (Fig. 4.14, top right). Although no vegetation could be seen at all in the areas where tsunami sediment had accumulated, we noticed that weeds were flourishing in a line around the paddies (Fig. 4.15). Upon close examination, these lines were found to be caterpillar tracks made by machines as they entered the paddies to remove debris such as pine trees. Comparison of the soil between the caterpillar tracks and the areas without vegetation revealed that the EC and available boron level were significantly lower in the caterpillar tracks areas (Table 4.4). This discovery served to validate the

Fig. 4.14 Surface of paddy field after removal of debris in Iwanoko, Soma City, in September 2011

Fig. 4.15 Weeds were found linearly only on caterpillar tracks

Table 4.4 Effects of soil mixing by movement of machines on soil chemical properties

Sampling site	EC mS/cm	Exchangeable Na$_2$O mg/100 g	Available B mg/kg
Area without vegetation	4.0	806	8.3
Area of caterpillar track	1.9	275	1.1

Fig. 4.16 Paddy field with mixed tsunami sediment and original soil (*left* of photograph) and unmixed paddy field (*right* of photograph)

salt removal method we had used in strawberry greenhouses, where the tsunami sediment was simply mixed with the original soil instead of being removed. Witnessing this phenomenon for themselves, the farmers decided to do the same in their paddy fields.

4.5.2 Soil Acidification Within the Expected Range

As shown in Fig. 4.16, tsunami sediment was mixed with the original soil in 60 ares of paddy fields on September 27, 2011, and mole drains were formed to improve water drainage toward the lower layer. Subsequently, rainwater alone was used for

Fig. 4.17 Application of converter slag using lime sower (April 2012)

salt removal. As a result, the EC, which serves as a guide for rice planting, dropped to 0.7 mS/cm by April 3, 2012 (232 days after soil mixing). However, the pyrite in the soil had gradually oxidized until the pH (H_2O) dropped to 3.8, so converter slag was applied on April 23 (Fig. 4.17) to neutralize the sulfuric acid.

4.5.3 Fields of Golden Rice Plants After a 2-Year Hiatus

In May 2012, rice of the variety known as Hitomebore was transplanted in three plots of 1.7 ha of paddies whose salt had been removed by the aforementioned method. No fertilizer was applied as we believed that the alkaline effect resulting from the rise in pH value caused by the application of converter slag would cause an increase in mineralization of organic nitrogen. The rice seedlings grew extremely well after transplanting, with no inconsistencies in their growth.

On September 26, a total of 10.7 t brown rice was harvested from the 1.7 ha of paddies (Fig. 4.18). The yield of brown rice harvested per hectare was 6.3 t, about 20 % higher than the amount before the disasters. Thus, we had managed to return the rice paddies to fields of gold after a 2-year hiatus.

Fig. 4.18 Paddy field of golden rice plants after a 2-year hiatus (September 2011)

4.5.4 The Highest Safety Standards for "Soma Revival Rice"

The city of Soma is located about 40 km from the Fukushima Daiichi nuclear power station. A radioactive cesium level of about 750 Bq/kg was detected in soil taken from paddies that produced this first crop of the rice that we were to call "Soma Revival Rice." However, not only was there no trace of radiation detected in the brown rice, but even the stems and leaves, which absorb radiocesium more easily than the grains, showed no trace of radiation at all. As cesium and potassium belong to the homologous elements, when a large quantity of potassium (exchangeable potassium of 25 mg/100 g or more) exists in the soil, it has been known to suppress the absorption of cesium by the rice. As the paddy soil mixed with tsunami sediment contained about 60 mg/100 g of exchangeable potassium, the process of mixing in the tsunami sediment had ended up suppressing radioactive cesium absorption by the rice.

The levels of harmful elements such as cadmium and arsenic measured in the rice itself were 0.01 mg/kg for cadmium and 0.04 mg/kg for arsenic (Table 4.5). In the case of cadmium, this was significantly lower than the 0.4 mg/kg that is the nationally stipulated limit in Japan. Although there is no limit set for arsenic, the level measured was only one-quarter of the 0.16 mg/kg average for rice produced in Japan. On the other hand, when levels of "healthy" minerals were compared with

Table 4.5 Cadmium, arsenic, and radioactive cesium concentration of "Soma Revival Rice"

Sample	Stem and leaf		Brown rice		Cd	As
	^{134}Cs	^{137}Cs	^{134}Cs	^{137}Cs		
Soma Revival Rice	N.D.	N.D.	N.D.	N.D.	0.02 mg/kg	0.04 mg/kg

Detection limit of radioactivity
    Stem and leaf: ^{134}Cs 9.3 Bq/kg ^{137}Cs 8.9 Bq/kg
    Brown rice: ^{134}Cs 1.7 Bq/kg ^{137}Cs 1.4 Bq/kg

Table 4.6 Mineral content of "Soma Revival Rice" (brown rice bases mg/100 g)

Sample	Na	Mg	P	K	Ca	Mn	Fe	Cu	Zn
Soma Revival Rice	1.6	144	365	266	8.6	2.4	2.3	0.3	2.1
Standard value[a]	1.0	110	290	230	9.0	2.1	2.1	0.3	1.8

[a]Standard tables of food composition in Japan

levels in the Standard Tables of Food Composition in Japan, the rice was found to contain higher levels of magnesium, phosphorous, and potassium (Table 4.6).

4.6   The Soma Project Expands Use of the Soma Method

4.6.1   The Project Launches

In total, 1,100 ha of farmland were affected by the tsunami in Soma, and only 140 ha were being farmed again by March 2013, the bulk of this being areas that suffered relatively minor damage. Of the farmland that was devastated by large volumes of debris and accumulated tsunami sediment, only a fraction had been restored and just 1.7 ha were the paddies we had rehabilitated using the Soma Method. We aimed, therefore, to restart farming on a larger scale by expanding the use of the Soma Method from an isolated 1.7-ha point on the map to a larger 50-ha area in 2013.

We therefore submitted a proposal and obtained the consent of the Soma City authority and JA Soma (the local agricultural cooperative) to launch the Soma Project. Subsequently, a representative of all three parties involved in the project requested the CEO of Japan's largest steelmaker, Nippon Steel & Sumitomo Metal Corporation, to help by supplying 450 t of converter slag at no cost, which the CEO promptly granted.

4.6.2   The Project Gets Under Way

On March 8, 2013, press conferences to announce the converter slag donation by Nippon Steel & Sumitomo Metal Corporation were held in Tokyo (Tekko Kaikan) and Soma's City Hall (Fig. 4.19).

Fig. 4.19 Press conferences of Soma Project in Soma City Hall (March 2013)

Fig. 4.20 Application of converter slag using lime sower by collaboration of farmers (April 2013)

Fig. 4.21 Rice planting was started in all the paddy fields of the Soma Project (May 2013)

Fig. 4.22 Gold-tinged ears of rice have started to droop down (August 2013)

In April, the Soma City Hall prepared a warehouse in Soma Port to accept a large amount of converter slag transported by 10-t trailers. At around the same time, the farmers started cultivating rice seedlings. Beginning April 19, a large tractor fitted with three lime sowers was used by the farmers to start scattering converter slag and mixing it into the soil (Fig. 4.20). Subsequently, the farmers flooded the paddy fields and wet tilled the land. Rice planting was then started in all the paddy fields from May onward (Fig. 4.21). In some paddies where the EC had been high, the farmers passed irrigation water through after the rice was planted. Thanks to these measures, the rice flourished without any problems, and as of August 30, gold-tinged ears of rice have started to droop down, ready to be harvested from mid-September onward (Fig. 4.22).

Chapter 5
Tsunami Damage to Farming Operations and the New Generation of Farmers and Farm Management

Yukio Shibuya, Takahiro Yamada, Nyamkhuu Batdelger, Puangkaew Lurhathaiopath, Gentaro Suzumura, and Toshiyuki Monma

Abstract In this chapter, we show a survey of farmers' inclination to resume farming and some actions of agricultural recovery that we carried out for the farmers where the city of Soma suffered from the tsunami disaster. Three agricultural corporations were established at the rice paddy area, and one agricultural corporation was established at the strawberry production area to resume farming in the tsunami disaster area. We show process of the establishment of these agricultural corporations and show the actual situation of the activity and a future problem.

Keywords Resume farming • Regional agricultural recovery associations • Tsunami damage • Agricultural corporation

Y. Shibuya (✉) • T. Yamada • G. Suzumura • T. Monma
Department of International Biobusiness Studies, Tokyo University of Agriculture,
1-1-1 Sakuragaoka, Setagaya-ku, Tokyo 156-8502, Japan
e-mail: y3shibuy@nodai.ac.jp

N. Batdelger
Business Management, National University of Mongolia,
Sukhbaatar Duureg, Ulaanbaatar 210646, Mongolia

P. Lurhathaiopath
Faculty of Life and Environmental Sciences, Tsukuba University,
1-1-1 Tennodai, Tsukuba City, Ibaraki Prefecture 305-8577, Japan

© The Author(s) 2015
T. Monma et al. (eds.), *Agricultural and Forestry Reconstruction After the Great East Japan Earthquake*, DOI 10.1007/978-4-431-55558-2_5

5.1 Damage to Key Farmers' Operations, Farmers' Inclination to Resume Farming, and Factors Inhibiting Resumption of Farming

5.1.1 Assessment of Damage to Farming Operations, and Survey Targeting Resumption of Farming

From June 12 to 15, and from June 24 to 29, 2011, we conducted an oral survey of 39 farming households affected by the tsunami in the city of Soma. We then conducted an aggregate analysis of 27 of these 39 households that grew rice as their main crop.

First, we investigated the proportion of paddy fields flooded by seawater and found that only 7 % of the total paddy area escaped flooding entirely, while the proportion flooded by 0–20 % was also low, at 4 %. On the other hand, 19 % of the paddy area was flooded by 60–80 % and 22 % was flooded by 80–99 %, while 37 % of the total area was 100 % flooded. Thus, despite differences in the level of damage among the survey respondents, we found that almost all of them had suffered some form of flooding damage. Next, we investigated the damage inflicted on the four main types of farming machinery, namely tractors, rice planters, combines, and dryers. The farmers were polarized in terms of the levels of damage sustained, with 59 % having suffered no damage at all to their machinery versus 30 % whose machinery was totally destroyed, and the remaining 11 % having suffered only partial damage.

5.1.2 Inclination to Farm in Future and Associated Factors

We asked the farmers about their inclination to continue farming, a key factor in considering the future of farming in the disaster zones. Five potential responses were allowed, with two of them—"Expand scale" and "Maintain current scale"— based on the assumption that farming would be continued. The other three responses were "Reduce scale," "Leave farming," or "I don't know." We also asked about the farmers' inclination to continue farming before the disasters, because we needed to compare their intentions before and after. The results are shown in the Total column in Table 5.1. After the disasters, the proportion of those who answered "Expand scale" or "Maintain current scale" was reduced, whereas the proportion of those who answered "Reduce scale," "Leave farming," or "I don't know" increased. From this data we note an overall decline in farmers' inclination to farm that can be attributed to the earthquake and tsunami disaster.

Upon analyzing the factors behind the decline in inclination to farm, we discovered that those who owned 5 ha or more of land showed some increase in pessimism compared to before the disaster, although none of these farmers replied "Leave farming" or "I don't know." On the other hand, there was increased pessimism after

Table 5.1   Changes in inclination to farm before and after the disasters

	All farmers $n = 27$		Farmers owned 5 ha or more $n = 11$		Farmers owned less than 5 ha $n = 16$	
	Before disaster	After disaster	Before disaster	After disaster	Before disaster	After disaster
Expand scale	5	4	4	3	1	1
Maintain current scale	20	11	7	5	13	6
Reduce scale	2	6	0	3	2	3
Leave farming	0	3	0	0	0	3
I don't know	0	3	0	0	0	3

Source: Cited from Shibuya et al. (2012)

the disaster in more than half of those who owned less than 5 ha of land and had wanted to maintain the current scale of their operations before the disaster.

We also analyzed the impact of damage to farmlands and farming machinery on inclination to farm. The results showed that, the greater the damage, the greater the impact both factors tended to have on inclination to farm. Furthermore, the difference in the correlation coefficients indicates that the extent of the damage to farming machinery had a bigger impact on the decline in inclination to farm than the damage to paddy fields typically caused by seawater flooding (Table 5.1). Although public aid could be expected for rehabilitating farmlands, farming machinery was as a rule bought and owned on an individual basis, and levels of damage varied. As a result, farmers whose machinery was damaged would need to replace all the lost machinery themselves, which would require substantial expenditure exceeding 10 million yen. And, if that was the case, they could expect no public funding. Meanwhile, a slump in the price of rice during the past few years is also thought to have accelerated the decline in inclination to farm.

However, there were still some large-scale farmers who suffered little damage. Such farmers continue to be inclined toward expanding their scale of operations, and this survey suggests that these bigger players might be able to take over the farmland of those who want to reduce their operations or leave the industry entirely.

5.2   Activities of Regional Agricultural Recovery Associations: Characteristics and Evaluation

5.2.1   Scheme to Subsidize Resumption of Farming Activities and Regional Agricultural Recovery Associations

The Ministry of Agriculture, Forestry and Fisheries created a scheme to help farmers start operating their farms again by paying a redevelopment subsidy of 35,000 yen for every 10 ares of paddy. The aim was to expedite resumption of farming and restore agriculture in the disaster zones. Under the scheme, the subsidy is paid via regional agricultural recovery associations to farmers affected by the disasters when they work together on activities to restart farming operations, such as straightforward garbage and debris removal, repair of ridges between rice paddies, repair of water conduits, and removal of salt.

5.2.2   Survey on Recovery Association Activities in Soma City and Issues Faced

We conducted an oral survey of 16 recovery associations in the city of Soma from November 12 to 21, 2011. The purpose was to establish what activities they were engaged in and the issues they faced, as well as to investigate what direction future recovery activities should take (Table 5.2).

Table 5.2 Overview of regional agricultural recovery associations in Soma

Number	Individual recovery association	Number of constituents (persons)	Area (a)
01	Tsukanobe	46	1,948.01
02	Ishigami	75	2,654.68
03	Niida	101	2,842.37
04	Haragama	146	2,796.03
05	Wada	105	4,798.04
06	Nakago	74	1,518.99
07	Motowarou	46	1,685.19
08	Minamiiibuchi	49	2,606.89
09	Iwanoko	153	9,923.76
10	Niida	123	13,629.58
11	Hodota	94	5,952.17
12	Nikkeshi	71	4,630.95
13	Kashiwazaki	148	10,278.05
14	Isobe	295	19,067.48
15	Koisobe	70	4,798.33
16	Yunukiyachida	20	1,238.1
17	Yunukitatemae	17	1,156.57
18	Fukushimaken-hamadori Farmers' joint association	40	12,178.77

Source: Batdelger et al. (2012)

5.2.3 Recovery Associations and Their Activities According to Level of Tsunami Damage

5.2.3.1 Classifying Tsunami Damage Level

Based on the research outcomes to date, we classified levels of tsunami damage from three perspectives—damage to farmland, damage to irrigation facilities such as water supply and drainage channels, and damage to farming machinery owned by the association members (Table 5.3).

Damage to farmland was classified according to how long the farmland had been flooded, the amount of building debris, the amount of debris from windbreaker forests, and accumulation of tsunami sediment, with two or three possible levels of damage defined for each of these factors.

Table 5.4 summarizes damage levels as assessed by the recovery associations according to the foregoing classifications. The numbers 01–17 represent the individual recovery associations. As the table shows, the reported levels of tsunami damage to farmlands, irrigation facilities, and farming machinery were diverse.

Next, two clusters were extracted using cluster analysis to classify the damage levels objectively (Fig. 5.1).

Table 5.3 Classification of damage sustained by regional agricultural recovery associations in Soma

Classification of damage		A	B	C
Damage to farmland	How long the farmland had been flooded	Less than 3 weeks (short period)	–	More than 3 weeks (long period)
	Amount of building debris	Nothing	Accumulation on some farmland	Far-reaching and extensive accumulation
	Debris from windbreaker forests	Nothing	–	A lot of accumulation
	Accumulation of tsunami sediment	Partially thin accumulation	Partially thick accumulation	Far-reaching extensive and thick accumulation
Damage to irrigation facilities		Functional maintenance	Slight and restored damage (only sediment accumulation)	Damage with difficult restoration
Damage to farming machinery		Nothing or several persons	Persons below a half	Persons more than a half

Source: Batdelger et al. (2012)

Notes: "A" in front shows damage smallness, as for "B"; "C" shows the damage situation in order of damage size among damage, and writers classified from the result of the interview
Moreover, the part currently described by "–" was classified according to two steps of damage situations of the size of damage

Table 5.4 Damage to recovery associations' farmlands and irrigation facilities and to members' farming machinery

Association no.		01	02	03	04	05	06	07	08	09
Damage to farmland	How long the farmland had been flooded	A	A	A	Unestablished association	A	C	A	A	C
	Amount of building debris	A	A	A		B	A	A	A	A
	Debris from windbreaker forests	A	A	A		C	C	A	A	C
	Accumulation of tsunami sediment	A	A	B		B	C	B	C	C
Damage to irrigation facilities		B	A	B		B	C	A	A	C
Damage to farming machinery		A	A	A		B	A	A	A	B
Association no.		10	11	12	13	14	15	16	17	18
Damage to farmland	How long the farmland had been flooded	C	A	C	C	C	C	A	A	Uninvestigated association
	Amount of building debris	A	A	C	C	C	C	A	B	
	Debris from windbreaker forests	C	A	C	C	C	C	C	A	
	Accumulation of tsunami sediment	C	B	C	C	C	C	B	B	
Damage to irrigation facilities		C	B	B	C	C	C	B	A	
Damage to farming machinery		C	A	A	A	C	C	A	A	

Source: Batdelger et al. (2012)
Notes: (1) "A" in front shows damage smallness, as for "B"; "C" shows the damage situation in order of damage size among damage, and writers classified from the result of the interview
(2) Association "04" was not established; Association "08" was not investigated

5.2.3.2 Features of Recovery Association Activities by Tsunami Damage Level

Table 5.5 summarizes features of the main activities carried out by the recovery associations according to the classification obtained from cluster analysis. As shown, the recovery associations that had suffered severe damage generally had joint operational structures, with priority given to cultivators when paying subsidies.

Specifically, the farmlands of associations in Cluster 1 Group I (12, 13, 14, 15) contained large volumes of post-tsunami glass and concrete fragments mixed into the soil, and radical measures to remove these were required. However, it was difficult for the associations to implement such measures on their own. At the time of the survey, therefore, only temporary fixes such as mowing of the ridges had been implemented and real agricultural reconstruction had not started. The Tokyo University of Agriculture subsequently addressed this issue by enlisting the help of a farming machinery manufacturer to conduct a demonstration on how to remove

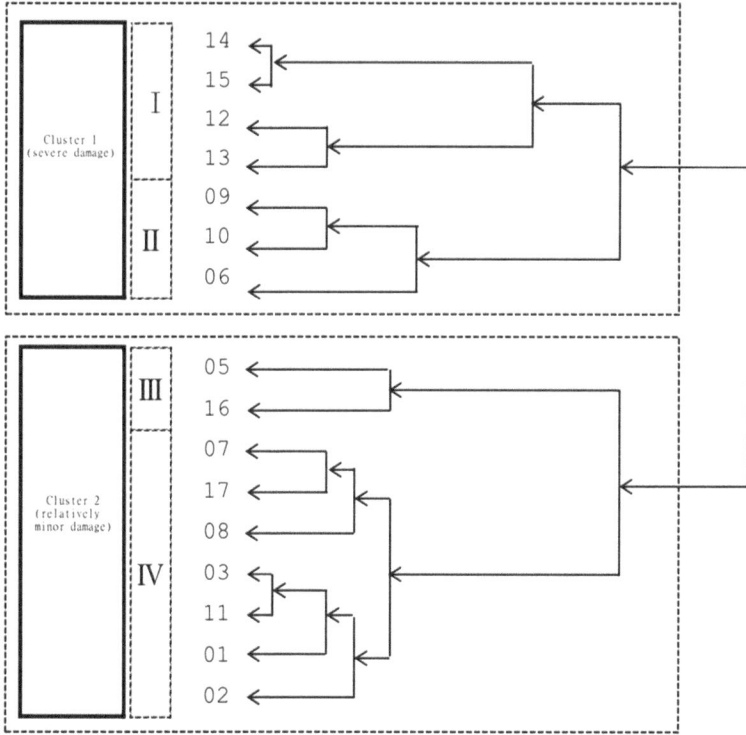

Fig. 5.1 Cluster analysis-based classification of regional agricultural recovery associations in Soma (From Batdelger et al. 2012)

fine debris buried in the soil. As a result, debris removal has now become one of the activities carried out by the associations themselves.

For associations classified under Cluster 1 Group II, whose farmlands contained no building debris (09, 10, 06), there was believed to be little need to remove topsoil. Moreover, we tried removing salt from some of Association 09's farmlands and found that the lack of glass and concrete fragments reduced the time and cost required for reconstruction significantly. In other words, it can be said that the subsidies granted under the government's scheme to help farmers resume farming are being used effectively for agricultural recovery and repair work.

Among recovery associations in areas that suffered relatively minor damage, we found that most of the repair work was undertaken at an individual level. We also managed to verify that wet tillage-based salt removal had already been carried out between three and five times by the associations or their members. In addition, most of the associations classified under Cluster 2 had undertaken repair work with the specific aim of planting rice by the following year. We found, therefore, that the subsidies were being used effectively for repair work to secure the incomes of the farmers affected by the disasters.

Table 5.5 Operational structures by classification, main recipients of subsidies, and future goals of activities

Recovery association no.		The main damage situations					Operational structures	Paying subsidies with priority	Activity target	Other
		Debris	Windbreaker forests	Tsunami sediment	Irrigation facilities	Farming machine				
Cluster 1 (severe damage)	I	14	C	C	C	C	C	Joint	Cultivators	A part of agricultural recovery	The restoration to farmland is difficult for the time being
		15	C	C	C	C	C	Joint	Cultivators+landowner	A part of agricultural recovery	–
		12	C	C	C	B	A	Joint	Cultivators+landowner	Recovery of the agriculture of 1/3	–
		13	C	C	C	C	A	Joint+individual	Cultivators	Nothing	–
	II	09	A	C	C	C	B	Joint	Cultivators (cultivators pay landowner)	Removing salt and Planting Paddy rice early	–
		10	A	C	C	C	C	Joint	Cultivators+landowner	Full-scale repair work is difficult until it determines the directivity of infrastructure improvement.	–
		06	A	C	C	C	A	Joint+individual	Landowner	–	–

Cluster 2 (relatively minor damage)											
III	05	B	C	B	B	B	Individual	Jobless people	Removing salt and Planting Paddy rice	–
IV	16	A	C	B	B	A	Individual	Cultivators	Removing salt and Planting Paddy rice	–
	07	A	A	B	A	A	Individual	Cultivators + landowner	Removing salt and Planting Paddy rice	–
	17	B	A	B	A	A	Individual	Cultivators	–	–
	08	A	A	C	A	A	Individual	Cultivators	Removing salt and Planting Paddy rice	Incorporation in the future
	03	A	A	B	B	A	Individual + joint	Cultivators	Removing salt and Planting Paddy rice	–
	11	A	A	B	B	A	Joint + individual + joint	Cultivators	Removing salt and Planting Paddy rice	Incorporation in the future
	01	A	A	A	B	A	Individual + joint	Cultivators + landowner	Removing salt and Planting Paddy rice	–
	02	A	A	A	A	A	Individual	Cultivators	Removing salt and Planting Paddy rice	–

Source: Batdelger et al. (2012)

"–" shows no answer

5.2.4   The Recovery Assistance Project: Challenges and Optimal Implementation Method

The key consideration in implementing a project to accelerate recovery is above all to deploy measures tailored to the specific level of damage. As seen in the past, uniform recovery assistance projects that ignore the diversity of the damage are not necessarily very effective. In addition, the establishment of recovery associations to overcome the harsh conditions imposed by these particular disasters belies the fact that farming via community-based organizations has not traditionally been the norm in Soma. Looking ahead, therefore, recovery association activities could be further developed to create new entities to help take responsibility for agricultural reconstruction. (Batdelger et al. 2012)

5.3   Impetus for the Formation of Agricultural Corporations and Process of Incorporation

5.3.1   Background of Agricultural Corporations in Soma

From the early stages after the disasters, the Soma City authority considered a policy of developing agricultural corporations. In February 2012, a restoration project focusing on soybean production in Soma was established with a 300 million yen grant from the Yamato Welfare Foundation. The aim of the project was to support the establishment of new agricultural corporations to produce, process, and sell soybeans in communities that were affected by the tsunami, such as Minami Iibuchi, Iwanoko, Niida, and Hodota. Specifically, the project provides farming machinery including large tractors of the 100-hp class required for soybean production.

5.3.2   Features of the Three Agricultural Corporations

Following is a brief overview of the three soybean corporations established in the city of Soma and the process leading up to their incorporation (see Table 5.6).

5.3.2.1   Iitoyo Farm LLC

Farmers in the Iitoyo area farmed about 2 ha on average on a self-sufficient basis and farmwork was almost never outsourced. The Soma City authority first suggested creating an agricultural corporation to the farmers in February 2012, and the company was incorporated within a very short period, at the beginning of April the same year.

Table 5.6 Soybean corporations established in Soma after the disasters

Corporation name	Iitoyo Farm LLC	Iwanoko Farm LLC	Agrifood Iibuchi LLC
Establishment area	Niida, Hodota	Iwanoko	Minamiiibuchi
Date of foundation	April 2, 2012	May 18, 2012	May 31, 2012
Disaster farmland area	196 ha (Niida 136 ha + Hodota 60 ha)	99 ha	26 ha
Recovery association number	217 houses (Niida 123 houses + Hodota 94 houses)	153 houses	49 houses
Total farmland area	240 ha	140 ha	40 ha
Directors	Niida 3 persons + Hodota 1 persons	4 persons	6 persons
Directors' feature	A full-time farmer is a center.	Part-time farm households, such as dried-seaweed aquaculture, are centers.	A horticultural farmer and a part-time farm household are centers.

One key feature of Iitoyo Farm is the size of the farming area that serves as its production base. In addition to the 190 ha that was damaged, there was another 50 ha that was undamaged, making a total of 240 ha. In addition, the directors of Iitoyo Farm are different from those of other corporations in that all four are full-time farmers. They cultivate 18 ha, 8 ha, 8 ha, and 2 ha, respectively, the first three directors clearly being large-scale farmers within the area. In addition, the director who cultivates 2 ha has long years of experience in housing sales, and could therefore play a key role in sales as the business expands in future.

As the board of Iitoyo Farm is composed of full-time farmers, it is not simply a community farm that aims to maintain its farmlands, but a corporation that aims to develop and expand its business continually in future. To ensure this continuity, the corporation needs to pay salaries on a par with other industries and secure talented staff who can be entrusted with the running of the business in future. The directors therefore urgently need to put the management of their business on a firm footing and raise revenue. If they continue to focus exclusively on crop production, as in the past, not only will work dry up in winter, but added value will also not increase. Therefore, the directors share the belief that a professional approach to agricultural production that includes processing and sales is the way forward.

5.3.2.2   Iwanoko Farm LLC

The Iwanoko district is situated in the northern part of the central reclaimed land next to Matsukawaura Lagoon. Consequently, many of the farmers cultivate edible seaweed in addition to farming. Currently, seaweed cultivation is the main business and there is no one who specializes in farming alone. The total farming area in the district measures 140 ha, with 100 ha producing rice and the remaining 40 ha producing soybeans. Before the disasters, a voluntary association called Iwanoko Farm undertook crop rotation-based production under a community farming structure. Soybean production amounted to a harvest of about 200 kg/10 ares. The local agricultural cooperative shipped the produce, which was of high quality and sold well.

The farm's managers currently have no plans to expand into a new business structure that can take on large investments and risks.

5.3.2.3 Agrifood Iibuchi LLC

The Minami Iibuchi district is located on the upper reaches of the reclaimed land. As the land under cultivation straddles the Route 6 bypass from east to west, the adverse effects of the tsunami were relatively minor compared to the damage to land cultivated by the other two corporations. The area between the bypass and the town, in particular, escaped almost unscathed. Only three or four farmers suffered damage to their machinery and only one person (one of the directors) left farming because of the tsunami. The total area of farmland in the district is small at 40 ha, with no farmers involved in full-time rice cultivation.

The farmers in this corporation's region are aging, and the board consists of only older and part-time farmers. They therefore need to secure successors to take over the business and to review the structure of the community's agriculture that has so far been composed of mainly part-time farmers.

5.4   Local Farmers' Expectations and Demands of the Agricultural Corporations

5.4.1   Survey Background and Method for Iitoyo Farm

Iitoyo Farm was established in April 2012, and by May it needed a specific business plan. That was the point at which the farm's managers realized that to prepare for full-scale resumption of farming the following year, they needed to understand (a) the extent to which they could expect the corporation to consolidate farmlands in the Iitoyo area (to indicate the corporation's likely scale of operations), (b) the potential for making the farmers more aware of the corporation and securing their collaboration, and (c) any specific requests the region's farmers may have.

From July 18 to August 8, 2012, we therefore conducted a questionnaire survey of farmers in the area where the Iitoyo Farm had been established. The number of questionnaires sent out was 208, and 90 replies were received, resulting in an effective response rate of 42.8 %.

5.4.2   Farming Before and After the Disasters and Changes in the Farmers' Inclination to Farm

Survey questions about how the paddies were cultivated and managed the year before the disaster revealed that the area of leased paddy was 80 ha and the area of paddy where all work was outsourced was 29 ha. Combining these two, the total leased or outsourced paddy area in the Iitoyo district was at least 110 ha before the disasters struck. Taking into account the 42.8 % effective response rate and the overall decline in farmers' inclination to continue farming their own land after the disasters (to be described later), it is possible that the amount of farmland either leased or outsourced could far outstrip that 110 ha in future.

We asked the farmers whether they were interested in entrusting their farmland to their community's agricultural corporation, the Iitoyo Farm. In response, 54 % replied that they would not mind doing so, and 10 % replied that it would depend on the services offered. Depending on how the corporation develops it services, therefore, this 10 % will probably also entrust their land to the corporation. Only 17 % clearly stated that they were against the idea. However, it is also likely that farmers who did not reply or were still undecided will find it difficult to cultivate their land themselves. As a consequence, consolidation of farmlands under the corporation is expected to increase significantly in future.

5.4.3   Management of the Iitoyo Farm by the Region's Farmers

The directors of Iitoyo Farm are four full-time farmers. Because they are all relatively young, ranging in age from 48 to 57, they are reasonably well able to cope with expansion of the area under cultivation. However, before the disaster, the two communities of Niida and Hodota that comprise the Iitoyo area already contained 110 ha of leased or outsourced land, and bearing in mind the number of farmers who are considering scaling down or leaving the industry as a result of the disaster, a total of 190 ha of damaged land may become available for leasing or outsourcing. If we exclude the 17 % of farmers who did not wish to entrust their farmland to the corporation, then 83 % of 190 ha, or approximately 160 ha of farmland, may be consolidated by the corporation. Furthermore, if the 50 ha of undamaged farmland were added to make the total available land 240 ha, then the area that may be consolidated would rise to 200 ha. With such a scale of operations, it would not be possible for the four directors to cultivate the land alone.

This being the case, collaboration among the farmers of the region is considered necessary. We therefore asked the farmers about how they intended to participate in the newly established agricultural corporation (Table 5.7).

These results suggest that there is a greater likelihood of consolidating a large farming area if it is assumed that local farmers will play a role within the corporation.

Table 5.7 Intended participation in the agricultural corporation

	Person himself/ herself	Parents, Spouse	Children
[Full-time] It is due to participation as a director.	5	–	0
[Full-time] It is easy to be based on operator work.	6	2	2
[Full-time] It is easy to be simple agricultural work and auxiliary work.	7	4	0
[Full-time] It is employment hope at a future director candidate.	2	1	0
[Temporary] It is easy to be operator work.	7	0	1
[Temporary] It is easy to be simple agricultural work and auxiliary work.	10	10	1
[Temporary] It is easy to be grass and water management work.	5	5	1
Not wish/Unknown/No answer	47	67	84
Sum total	89	89	89

Source: Cited from Shibuya et al. (2013)

5.4.4 Agricultural Corporations in the Disaster Zones: Potential and Expectations

Large-scale agricultural corporations focusing on rice cultivation can be seen across the country, but none of them grew big overnight. Instead, they expanded to their current scale gradually as they accumulated experience and knowledge. In contrast, the agricultural corporations established in the disaster zones face unprecedented challenges as large areas exceeding 100 ha have to be farmed even during the 2- to 3-year period when the farmland and irrigation infrastructure are being restored. For this reason, business management capabilities need to be enhanced.

The success of the agricultural corporations established in the city of Soma is expected to serve as a model not only for the reconstruction of rice paddies affected by the Great East Japan Earthquake but also for resolving issues faced by the rice cultivation industry in Japan as a whole as a consequence of its aging population.

5.5 The Strawberry Farm Corporation: Characteristics of Activities and Future Strategy

5.5.1 Overview of the Wada District's Strawberry-Producing Area

The Wada district of the city of Soma lies to the west of Matsukawaura Lagoon and is located relatively close to the tourism and hotel district centered on Matsukawaura. Strawberry production started a long time ago in this area, and the Wada strawberry

Table 5.8 Overview of Wada strawberry farm

Corporation name	Limited liability company: "Wada strawberry Farm"
Date of foundation	May 11, 2012
Capital	1 million yen
Corporate directors	7 persons
Institution scale	Cultivation house: Two houses of 650 tsubos, one house of 450 tsubos (all are an elevated culturing apparatus)
	Seedling raising house: Four vinyl houses of 30 meters, four vinyl houses of 40 meters
Business name	Recovery grant
Amount of money	240 million yen (the part in the Heisei 24 fiscal year)

association was established in 1988. The members of the association each managed their own operations, and the number of members before the disaster was only 13 in this small-scale strawberry-producing area.

5.5.2 Events Leading to the Establishment of the Corporation

Among the 13 members, 7 were forced to cease operating as a result of the tsunami caused by the Great East Japan Earthquake, and only 6 sustained no damage to their greenhouses. As revenue was derived from strawberry picking and direct sales, business was dependent on visitors coming. But if the greenhouse area available was reduced, the district would be unable to fully cater to its visitors, losing its appeal as a strawberry-producing area. As a consequence, any potential reduction in the number of strawberry farmers was a problem that affected the whole association. The farmers therefore consulted the Soma City authority and sought national recovery aid to incorporate a limited liability company (Table 5.8).

5.5.3 Overview of Events Leading to Construction of the Greenhouses and Their Management

The greenhouses in the Wada strawberry farm were all built using a recovery grant (Fig.5.2). Following discussion, the members decided that, instead of owning the greenhouses as individual assets, they would build and operate them as assets of the limited liability company. Behind this decision was the issue of succession. If a sturdy greenhouse is built, it can be operated continuously for several decades. However, most of the members did not have any confirmed successors and it was possible that, despite having greenhouses, there could be no one to take them over. They therefore decided that building and operating the greenhouses under the auspices of the association would lead to their continued and effective use.

Fig. 5.2 Linked strawberry greenhouses (*left*) and seedling greenhouse (*right*) constructed using a recovery grant

5.5.4 Discussion of Incorporation Initiative in the Wada District

Despite administering a small area with only 13 farming households, the Wada strawberry association has managed to develop during its 25 years. Using a national recovery grant, the association's members set themselves up to repair their production facilities. Considering the future need for farmers and overall management of the business, they also took the opportunity to switch from an individual to a corporate management structure.

One common worry faced by farmers in areas such as this that have been producing agricultural products for 25 years, or almost a generation, is the issue of succession. In the Wada district, the crisis brought on by the disasters provided the impetus for the farmers to discover incorporation as a means to resolve this perennial issue. Indeed, this example of independent operators joining together merits attention as a means to solve the issue of succession among farmers, whose businesses are said to be the most difficult of all businesses to consolidate. The farmers' choice on this occasion reflected the fact that it was easy to reach a consensus within the Wada district, a small area with only 13 farming households. Whether this decision will turn out to be a blessing depends largely on the future efforts of the members of the limited liability company.

5.6 Recovery Based on Agricultural Corporations: Future Direction and Issues

5.6.1 Agriculture in Tsunami-Stricken Areas of Soma and Its Features

Within the city of Soma, the areas affected by the disasters were mainly paddy fields on reclaimed land that had continued to enjoy favorable conditions since the Edo Period (1603–1868). The farmers' landholdings were larger than was the norm in Japan, averaging about 2 ha in Niida district, for example. Even now that many farmers farm only part-time, self-sufficient, individual management is usual, and at the time of the disasters community farming and agricultural corporations were almost unknown.

However, a long-term slump in the price of rice meant that conditions were getting tougher year after year. Amid such a harsh business environment, agriculture had become a way to maintain the family property rather than a means to acquire income, and as such it was inevitably sustained mainly by senior citizens. For many years before the disasters it had been the norm for potential successors among the young to find work outside of agriculture, and some farmers had actually started leaving the industry for lack of a successor. It can therefore be said that, even before the disasters, agriculture in Soma was nearing crisis point in terms of both revenue and people to take over. This problem is not limited to the city of Soma, but is common to all agricultural regions throughout Japan.

5.6.2 Recovery Based on Incorporation: Future Direction and Strategy

Thus, paddy field rice cultivation in Soma was barely surviving. Then the farmlands serving as a common production base for the region were severely damaged as a result of the tsunami, along with individual assets such as farming machinery. As far as self-sufficient rice cultivation was concerned, the damage to machinery in particular became a key factor in farmers deciding to give up farming.

Naturally, the city needs to avoid a situation whereby the national budget is used to restore farmland, facilities, and farming machinery to their original state as part of the disaster recovery effort, only for these assets to fall into disuse several years later. However, simply returning to the previous self-sufficient model risks wasting the recovery budget because of the continual crisis brought about by farmers retiring.

After the disasters, the farmers' attitudes to farming changed because of such concerns about succession and the damage their farms and machinery had sustained. The city authority reacted to these changes in attitude by promoting incorporation both inside and outside the city from the early stages of recovery planning. This approach set the scene for the city authority, which applied for the Yamato Welfare Foundation's reconstruction and recovery fund, to stipulate incorporation as a condition when farmers applied to lease farming machinery. However, the city authority did more than simply lay down conditions; it deserves mention for also assisting with paperwork, providing advice, and building consensus in a timely manner during the incorporation process. In the Wada district, the city also advanced the cause of incorporation by building the latest strawberry greenhouses using elevated hydroponic cultivation with the help of a government recovery grant. Incorporation offers huge potential for improving a farming business economically in terms of operating efficiency and profitability because of the greater economy of scale. It also offers huge potential for improvement from a human resources point of view by providing a means of ensuring succession.

Management resources can generally be said to consist of people, equipment, and money. Equipment (large farming machinery and the latest greenhouses) acquired through recovery projects can serve as a means of bringing in people (successors) and money (profitability) to furnish all three resources. With these three elements in place, the city of Soma can be said to have taken a strategic approach to establishing the basic conditions for its agriculture to grow over the long term. The strategy aims to resolve local agricultural management issues by using equipment as the means to induce people- and money-related changes. Besides serving as a model for other regions recovering from the disasters, such a strategy can be said to represent a breakthrough for Japan's agricultural industry in general, which is facing the same issues.

5.6.3 Agricultural Recovery Based on Incorporation: Issues

The crisis caused by the unprecedented tsunami damage can be used as a catalyst to restructure the agricultural industry, and incorporation can be described as key to that strategy. The strategy is just getting under way in the city of Soma, but there are issues involved. In terms of the three management resources, simply building an environment where people (successors) can be found is not sufficient. Perhaps the biggest issue is whether Soma's existing farmers within the new corporations can make use of the equipment (large farming machinery and the latest greenhouses) that the corporations and the recovery process have made available to them. Only by using that equipment to realize greater economy of scale, systematization, and efficiency can they achieve the desired management outcome of generating money (increased profits).

Farmers therefore need to discard the mindset cultivated during the era of self-sufficiency and acquire the new mindset and management skills of corporate managers.

References

Batdelger N, Yamada T, Suzumura G, Shibuya Y, Lurhathaiopath P, Monma T (2012) Actual situation and evaluation of reconstruction union activities in tsunami-damaged area. J Rural Econ 2012(Special Issue):192–198 (in Japanese)

Shibuya Y, Yamada T, Batdelger N, Lurhathaiopath P, Niitsuma T, Usuki M, Monma T (2012) A study of intention to keep on farming and its factors on the farmers suffered from the Great East Japan Earthquake. Jpn J Farm Manag 50(2):66–71 (in Japanese)

Shibuya Y, Yamada T, Monma T (2013) Trends and subjects on incorporation of agriculture at tsunami stricken area: a case study in Soma, Fukushima. Jpn J Farm Manag 50(4):87–92 (in Japanese)

Chapter 6
Presenting a Model for the Revival of Rural Communities in Japan's Disaster Zones

Shigeyuki Miyabayashi, Yasushi Takeuchi, Hiromu Okazawa, Tomonori Fujikawa, and Yutaka Sasaki

Abstract During our research for the Great East Japan Earthquake Recovery Assistance Project, we focused on the city of Iwanuma in Miyagi Prefecture, where we held workshops with farmers to discuss how the revival of agriculture in the area should proceed. The feedback from these workshops was subsequently reflected in our revival plan. However, scientific data on how the ground had been altered by the earthquake and tsunami was also required to hasten the area's reconstruction and revival. To this end, we used a three-dimensional (3-D) laser measurement system with MMS (Mobile Mapping System) to reveal the changes in the topography of the survey district. Our results suggest that the MMS system can be applied across a diverse range of fields related to community development, such as improving emergency evacuation routes and road infrastructure, devising town planning models, or for agricultural production forecasting. In future, we plan to survey a wider area, and to use our findings as a baseline ground map for actual restoration work as we conduct further research into how these baseline data can be applied to community development.

Keywords Workshop • Revival plan • 3-D laser measurement system • MMS (Mobile Mapping System) • Community development

S. Miyabayashi (✉)
Department of Forest Science, Tokyo University Agriculture,
1-1-1 Sakuragaoka, Setagaya-ku, Tokyo 156-8502, Japan
e-mail: miyas@nodai.ac.jp

Y. Takeuchi • H. Okazawa • T. Fujikawa • Y. Sasaki
Department of Agricultural Engineering, Tokyo University Agriculture,
1-1-1 Sakuragaoka, Setagaya-ku, Tokyo 156-8502, Japan

© The Author(s) 2015
T. Monma et al. (eds.), *Agricultural and Forestry Reconstruction*
After the Great East Japan Earthquake, DOI 10.1007/978-4-431-55558-2_6

6.1   Introduction

The Faculty of Regional Environment Science and its three departments (the Department of Forest Science, the Department of Bioproduction and Environment Engineering, and the Department of Landscape Architecture Science) each utilize their areas of specialty to establish community development theories. We at the faculty consider watersheds to be units of community, and we hold symposiums to discuss frameworks for community development within watersheds. Such frameworks are structured around three key themes: human development (education theory), physical development (planning and technology theory), and conceptual development (policies and economic theory). We believe that our frameworks for community development based on these three themes can be used in efforts to revive rural communities affected by the Great East Japan Earthquake and to help them function effectively again. It was within this context, therefore, that we chose a study of community development in the coastal villages of Fukushima Prefecture's Soso area as the topic for our faculty project for the 2011 academic year, and subsequently worked to conduct research on community development.

Our research produced the following findings: (a) the functioning of drainage channels in this coastal watershed area had been seriously impaired as a result of severe land subsidence, making agricultural land susceptible to flood damage; (b) the massive deposits of sediment washed across farmland by the tsunami were unsuitable for use as agricultural soil or in building materials, and were best used as embankment material after decontamination treatment; and (c) the backfill sand in many village drainpipes had suffered some liquefaction, leaving numerous sites susceptible to future damage such as road cave-ins. Following on from our research we held agricultural workshops in the city of Iwanuma in Miyagi Prefecture, developing a long-term vision for agriculture, and presenting our suggestions to the Iwanuma Agricultural Recovery Committee.

Our research entailed conducting a baseline study aimed at facilitating further reconstruction and revival, and combining data obtained from this baseline study with 3-D digital maps of the area in collaborative research involving industry, academia, and government [topographic maps that include buildings can be created using a vehicle-mounted MMS (Mobile Mapping System) consisting of a digital camera and a 3-D laser measurement device; the level of accuracy is 1/500]. Thus, we aimed to study how rural communities should be revived by simulating local landscapes, the effects of restoration initiatives, and other scenarios.

6.2 Agricultural Revival and Workshops in Iwanuma, Miyagi Prefecture

6.2.1 Agricultural Revival Workshops

The Great East Japan Earthquake and the tsunami that followed claimed 150 lives in the city of Iwanuma (including 2 residents still missing), and damaged a total of 2,766 houses (699 completely destroyed, 421 mostly destroyed, 636 partially destroyed, and 1,010 partially damaged). In addition, the tsunami destroyed 9.9 km of seawall along the coast, flooding the inland area and causing destruction across the farmland and the industrial zone (approximately 230 businesses were affected). Overall, the area suffered more than 4 billion yen worth of damage.

The disasters wreaked the greatest destruction in agriculture, the city's main industry. In addition to despoiling about 1,240 ha of agricultural land, the tsunami also caused significant damage to agriculture-related infrastructure and farm machinery. The agricultural production systems that had been operated by individual farmers or collectives collapsed, as did the communities that had supported them.

To revive local agriculture in these circumstances, the farmers urgently needed to hold detailed discussions about the future of agriculture in the area. Such discussions would enable them to establish a range of strategies based on communal consensus and to work toward recommencing sustainable farming that could offer hope to the community.

We therefore decided to hold workshops as a means to discuss how to work toward resuming commercial agriculture in line with the community's visions for its future (Fig. 6.1). The workshops focused on postdisaster use of farmland, agricultural production, and how to cultivate organizations and individuals to lead the community going forward. Three workshops were held, on December 22, 2011, and on January 21 and 22, 2012.

Fig. 6.1 (*Left*) Schematic representation for the future of Iwanuma City's agriculture obtained from the results of the workshop (*right*)

6.2.2   Overview of Results of Agricultural Workshops

Before the Great East Japan Earthquake and tsunami, farming in Iwanuma already reflected the grim current state of Japanese agriculture as a whole. In addition to depressed prices for agricultural products and the aging of the farming community, the area also faced a lack of young successors. Then the massive tsunami generated by the earthquake brought new challenges, such as land subsidence and salt damage, as well as the loss of much of the city's production base, including farmland, agricultural machinery, and other capital equipment. The disasters left the farmers facing a whole range of additional long-term issues: lack of funding to restart farming, the collapse of the farming community, the uncertainty of the path toward agricultural revival, the fear of reputation-based economic damage, and the growing shortage of successors and other individuals to take roles of responsibility. In addition to the long-term concerns, however, the new challenges also highlighted other very immediate, practical concerns: How would the farmers farm in future? When would they be able to resume farming? How would they secure new funding to purchase the necessary machinery? Would the agricultural community really be able to recover?

To develop a revival plan that addressed the needs and wishes of local residents, we needed to ascertain what the conditions were in the area following the earthquake and ensuing disasters, and in particular how the ground, roads, water systems, and other physical features had changed. This realization prompted us to conduct a fact-finding survey using the MMS observation device outlined next.

6.3   Topographic Analysis and Community Development Using an MMS (Mobile Mapping System)

6.3.1   What Is MMS?

MMS is a three-dimensional (3-D) laser measurement system (based on technology developed by NTT Geospace Corporation) that uses a vehicle-mounted digital camera and 3-D laser measurement device to create 3-D digital maps of an area's topography, including buildings. The survey vehicle is fitted with an omnidirectional camera that captures 900 images per square meter (at 40 kph driving speed). These images are then processed by laser classification (class 1), allowing creation of 3-D maps (to 1/500 accuracy) (Fig. 6.2).

System type	STREET MAPPER		
Maker	IGI mbH		
System specification	Omni directional camera	Lady Bug 3 (PointGrey Inc)	
		6high quality 1600x1200pixels camera	
	Laser scanner	VQ250 (Riegl Ltd.)	
		Laser product classification	: Class 1
		Laser Pulse Repetition Rate	: ~300kHz
		Max. Effective Measurement Rate	: up to 300m (depending on Pulse Measurement Rate)
		Accuracy	: 10mm
		Resolutions	: 900 points/(at speed of 40km/h)

Fig. 6.2 MMS-mounted vehicle and specification of MMS

6.3.2   Details of Survey Process

We began by creating 3-D base maps using aerial photographs of the district as it was before the disasters, and postdisaster MMS data, to ascertain the changes in topography following the earthquake and tsunami in the west of the agricultural recovery zone in which the study was conducted. First, we used aerial photographs taken in October 2010 to create 1/2,500 scale 3-D topographic base data for the target district. Next, we used an MMS-equipped vehicle to take measurements of the area, and used the data obtained to construct 1/500–1/1,000 scale 3-D topographic base data. These data allowed us to identify how the earthquake and tsunami had altered the ground in the district.

6.3.3   Survey Results (Changes in Ground Topography)

Ground displacement in both the horizontal and vertical directions was observed in the survey district, with horizontal displacement of 2.72 m. Displacement of 2–4 m from the southeast to the southeast and northeast directions was also observed. Further, the vertical displacement (ground subsidence) at each point was compared using data from pre-earthquake aerial photographs and post-earthquake MMS data at points where the same features (white lines, etc.) could be clearly identified. This analysis found subsidence ranging from 20 to 70 cm (Fig. 6.3).

6.4   Summary

During our research for the Great East Japan Earthquake Recovery Assistance Project, we focused on the city of Iwanuma in Miyagi Prefecture, where we held workshops with farmers to discuss how the revival of agriculture in the area should

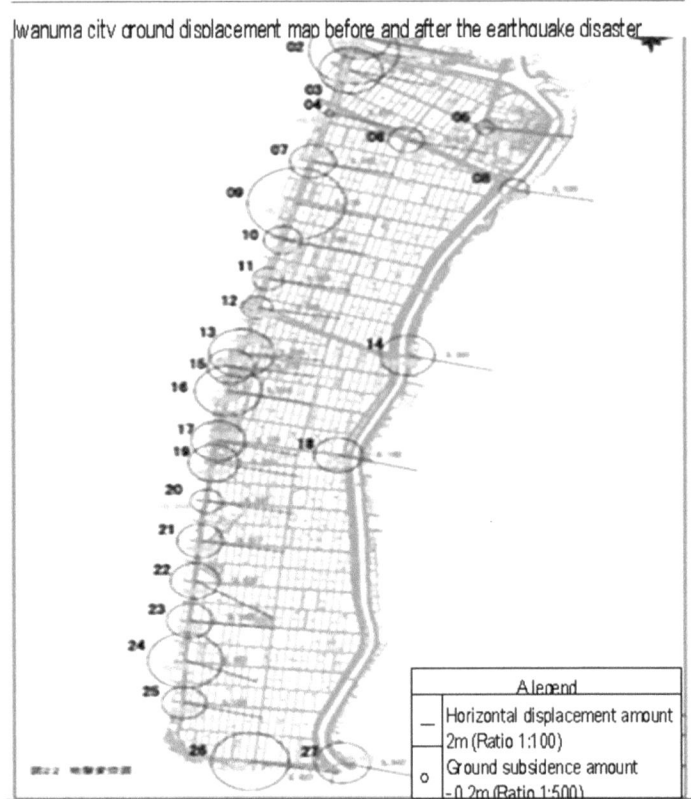

Fig. 6.3 Overview of changes of the ground in the surveyed area

proceed. The feedback from these workshops was subsequently reflected in our revival plan. However, scientific data on how the ground had been altered by the earthquake and tsunami were also required to hasten the area's reconstruction and revival. To this end, we collaborated with both government (the Iwanuma city authority) and industry in the form of a joint research project with NTT Geospace Corporation, Sankei Engineering Co., Ltd., and Asco Co., Ltd., using a 3-D laser measurement system with MMS to reveal the changes in the topography of the survey district.

Given the topographic changes, the base maps we compiled should be used for reference to plan water conduits and farm roads when maintaining or improving agricultural fields. The maps highlighted several aspects requiring caution, most notably the need for care when planning water conduits as the land is below sea level at certain points.

We were also able to confirm the utility of the MMS 3-D laser measurement device (and the MMS system as a whole) for creating agricultural and community revival plans following disasters. Furthermore, our results suggest that the MMS system can be applied across a diverse range of fields related to community development, such as improving emergency evacuation routes and road infrastructure, devising town planning models, or for agricultural production forecasting. In future, we plan to survey a wider area, and to use our findings as a baseline ground map for actual restoration work as we conduct further research into how these baseline data can be applied to community development.

We believe that ongoing, effective use of the data obtained from this type of collaborative project involving industry, academia, and government will prove beneficial in developing a model for simulating revival plans for rural communities.

Chapter 7
Contributing to Restoration of Tidal Flats in Miyagi Prefecture's Moune Bay Following the Great East Japan Earthquake and Tsunami

Susumu Chiba, Takeshi Sonoda, Makoto Hatakeyama, and Katsuhide Yokoyama

Abstract The Great East Japan Earthquake destroyed many artificial structures on the coast of northeastern Japan, and as a result it restored wetlands in many places. To conserve these wetlands, we started to estimate the ecological and economic values of a restored mudflat in Moune Bay, Miyagi Prefecture, Japan. Our tasks could be classified into three main categories: (1) establish the distribution of commercially important clams; (2) investigate the dynamics of the benthic community inhabiting the tidal flats and the floor of the bay; and (3) create a system so that monitoring of the tidal flat environment—including the two tasks mentioned—could be performed by nonspecialists. We hope that our actions will help to stimulate discussion about how Japan's coastal areas should be used.

Keywords Asari • Moune Bay • Tidal flat • Tsunami • Wetland

7.1 The Impact of the Great East Japan Earthquake and Tsunami on the Sanriku Coast

The coast of the northern Tohoku region, stretching from the city of Ishinomaki in Miyagi Prefecture to the city of Hachinohe in Aomori Prefecture, is a "ria" coast, consisting of a collection of bays of varying sizes that back onto steep mountains.

S. Chiba (✉) • T. Sonoda
Faculty of Bioindustry, Tokyo University of Agriculture, Abashiri, Hokkaido 099-2493, Japan
e-mail: s2chiba@bioindustry.nodai.ac.jp

M. Hatakeyama
Mori wa Umi no Koibito (Non Profit Organization), Kesennuma, Miyagi 988-0527, Japan

K. Yokoyama
Department of Civil and Environmental Engineering, Tokyo Metropolitan University, Hachioji, Tokyo 192-0397, Japan

© The Author(s) 2015
T. Monma et al. (eds.), *Agricultural and Forestry Reconstruction After the Great East Japan Earthquake*, DOI 10.1007/978-4-431-55558-2_7

These inlets protected by mountains proved to be good harbors, and many small settlements formed on what flat land there was around the bays. However, it was these very topographic features of the rias that amplified the power of the massive tsunami on March 11, 2011, extending the damage right up to the high ground. The giant wave devastated most settlements on the Sanriku coast, claiming the lives of many residents. Furthermore, once the tsunami had subsided it became apparent that the ground had sunk by tens of centimeters. This subsidence was enough to further shrink the already scarce area of flat land available in the Sanriku bays, and some areas where people had resided were now part of the sea.

Even two and a half years later, the pace of recovery is slow and little progress has been made. Many Sanriku coast residents affected by the disaster still live in temporary housing, partly because many cannot return to their ruined homes, but also reflecting the residents' uncertainty over whether it is wise to return to the area. Amidst such uncertainty, the local authorities made a unilateral decision to build seawalls 10 to 15 m high around much of the Sanriku coast. These giant walls might perhaps give the residents some peace of mind, but at the same time will deprive them of the view from the land and the breeze from the wind that once swept across the sea. The fishermen will likely have to scale these walls each day to check the sea conditions. And will the tourists who visit the Sanriku region for its beautiful views still leave satisfied? Although nothing is more important for a society than protecting life, the plan to obscure the Sanriku coast with seawalls leaves plenty of scope for debate.

7.2 Moune Bay and the NPO Mori wa Umi no Koibito

Moune Bay is a small inlet typical of those in the Sanriku region, located east of the city of Kesennuma in Miyagi Prefecture (Fig. 7.1). The bay settlement was unavoidably engulfed by the massive tsunami, and the land also subsided 70 cm, turning much of the area into tidal flats, or wetlands filled with halophytic plants. One can hardly begin to imagine the sense of loss and despair felt by the residents who lost their home, a place to which to return.

In the midst of the post-tsunami chaos, the NPO Mori wa Umi no Koibito (translated literally as "the forest is longing for the sea") has led discussion on land usage in the Moune district, with a view to restoring the tidal flats. This NPO is led by Shigeatsu Hatakeyama, a fisheries operator who was among the world's first proponents of tree-planting initiatives to protect the ocean. Hatakeyama's book *Mori wa umi no koibito* (published by Bungeishunju Ltd.) is well-known in Japan. Thanks to his book and activities, the Moune district had been a focus of attention even before the disaster, hosting numerous visitors from both within Japan and overseas, and many of its residents were highly conscious of the relationship between humans and nature. Even with this background, however, it was only natural that after the tsunami the residents were torn between reclaiming as much of the former land as they could or starting completely afresh. It was not a decision to be taken lightly. Eventually, with residents in two minds, the deciding factor proved to be the discovery of tiny *asari* (Japanese littleneck) clams, measuring less than 5 mm, on what had previously been a road.

Fig. 7.1 Moune Bay. The bay settlement was engulfed by the tsunami and the land subsided. (Photograph courtesy of IDEA Consultants, Inc.)

The *asari* clam is a bivalve that has been used for food in Japan since ancient times. In recent years, however, the quantity caught across the country has plunged. Although there are several reasons for this drastic decline in *asari* numbers, the shift in the way humans have used coastal areas in recent years is a major factor. The human desire to utilize land to its limits, combined with a rapid advance in civil engineering technology, has seen solid land extended to wetland and shore areas, replacing the ecotone—the vague boundary that separates land from sea—with a clear-cut demarcation. This demarcation has made it possible to expand the territory available for human use, providing additional habitable area even in the Sanriku region, where flat land is scarce. However, it was in this vague boundary between land and sea that natural resources were most abundant. The *asari* clam was one of these.

In recent years, the Japanese coastline has paid the price for this short-sighted pursuit of profit, with many of the natural assets taken for granted in the past quietly starting to disappear. The Great East Japan Earthquake was a terrible disaster that caused unspeakable loss—that fact is beyond debate. However, it is also an unfortunate truth that the resulting breakdown of the boundary between land and sea allowed us to once again catch a glimpse of nature's fading bounty. It was in fact not until the 1960s that the coast of Moune Bay was filled in to its pre-earthquake form; as recently as 50 years ago the area had originally been wetland, consisting mainly of tidal flats. So perhaps the discovery of *asari* clams on a road that the earthquake had turned into a tidal flat made Moune residents nostalgic for the old days, or perhaps it gave them new hope for the future. Either way, the discovery of the clams proved the impetus for a petition presented to Kesennuma's city authority in June

2012, stating the residents' intent to live in harmony with the ocean, without relying on seawalls. At the time, this was the only such decision made by residents on the Sanriku coast.

7.3 The Role of the Tokyo University of Agriculture Team

The discovery of *asari* clams in the area gave the Tokyo University of Agriculture group an opportunity to cooperate in the restoration of tidal flats in the Moune bay. Our tasks could be classified into three main categories: (1) establish the distribution of *asari* clams within the Moune area; (2) investigate the organisms inhabiting the tidal flats and the floor of the bay beyond, and how they changed over time (in other words, study the dynamics of the benthic community); and (3) create a system so that monitoring of the tidal flat environment—including the two tasks above—could be performed by nonspecialists. Of these three tasks, our responsibility for the third was particularly large.

As part of its environmental education program, the NPO Mori wa Umi no Koibito holds courses attended by many local elementary and junior high school students. If we could work with the NPO to develop our initiatives into a monitoring-based education program for children, it would enable monitoring to be performed on an ongoing basis. And it would be a wonderful added bonus if the experience of monitoring as children resulted in an increased number of adults with an appreciation of all that they owed to their natural surroundings.

7.4 *Asari* Clams

We selected a survey site in the back of the Nishi-moune bay, consisting of a 3,000 m^2 area enclosed by two small rivers, the Nishi-moune River and the Higashi-moune River (Fig. 7.2). The area previously contained a concrete-covered coast

Fig. 7.2 A monitoring area of tidal flats at high tide (*left*) and low tide (*right*)

Fig. 7.3 Asari clams and other benthos collected from quadrat of 20×20 cm

wall that served as both an embankment and a road. Located behind this had been wind-blocking trees, houses, and farmland, but the 70-cm ground subsidence means that the area is now completely submerged at high tide. We have visited the site once every 6 weeks since the study began in May 2012, continuing the same survey work.

From our investigation results it became evident that the *asari* clams discovered on the tidal flat that had formerly been land were not a chance finding: they were clearly inhabiting the area (Fig. 7.3). Most of the clams collected in 2012 were still small, measuring less than 10 mm in length. As an overall estimate taking into consideration factors such as the frequency of shell length and the number and width of ring patterns on shells, the clams measuring less than 10 mm were believed to be almost exactly 1 year old. This finding suggests that the spat (young shellfish) emerged after the March 2011 earthquake and subsequently inhabited the tidal flat.

The concentration of juveniles of the *asari* clams exceeded 6,500 individuals per square meter in highly populated areas, and exceeded 4,000 individuals on average in the tidal flat that had formed beside the former embankment road. Most surprisingly, this concentration was as high as any of the major *asari* clam production sites in Japan. At present, however, this outstanding level of production is limited to one part of the tidal flat, and it is an environment that, having once been dry land, is not now ideally suited to either humans or marine life. If anything, we should be impressed by the fact that the area was populated with so many clams despite the poor environment. Indeed, the discovery of the clams is of major significance in considering the value of reviving the Moune tidal flats.

7.5 Other Benthos

Asari clams are a symbol of the revival of the Moune tidal flats, and an important species from the fishing industry's perspective because of their market value. However, from an ecological perspective *asari* clams are merely one species in the tidal flat ecosystem. As the ecosystem comprises a combination of mutual interactions between various organisms and the environment, it is impossible to gain an understanding of it from studying a single species in isolation. An essential research task, therefore, is to study the dynamics of other benthic communities, even if the ultimate objective remains *asari* clam cultivation.

In addition to *asari* clams, the existence of 16 species of shellfish, 4 species of decapod crustacean such as shrimps and crabs, and 3 species of polychaetes such as ragworms has been confirmed at present. When we began the study, most of these organisms were significantly smaller than they would be when fully grown. Similar to the *asari* clams, they are therefore believed to have entered the newly formed tidal flat at a later date, rather than being washed inland by the tsunami. With each subsequent study, as time passed following the disaster, the number of species present gradually increased, suggesting that the tidal flat is steadily becoming closer to functioning normally again.

7.6 Development of Monitoring Techniques

We are currently working to develop monitoring techniques in the hope that local residents—particularly elementary and junior high school students—will eventually take over the task of monitoring the tidal flats (Fig. 7.4). Above all these

Fig. 7.4 Development of accurate and simple monitoring techniques for local residents

techniques need to be simple, so that anyone can perform them. At the same time, however, we are aiming to collect data that will actually be useful 10 or 50 years in the future.

The most difficult task for monitors is classification of the benthos. With the exception of *asari* clams, nonspecialists are unable to identify by name many of the benthic species. Although the difference between shrimps, crabs, and shellfish is understood, at levels above this many organisms appear similar to the untrained eye. To address this issue, we began by taking photographs of each benthic organism collected and are currently working to create illustrations annotated with their characteristics.

The biggest challenge during this task is likely to be capturing children's interest: if they do not find the monitoring process interesting, they will not keep it up. As it is important that we find the facts we need, we are reviewing the monitoring methods through trial and error so that we do not end up with an ineffectual plan created from an adult perspective.

7.7   Future Challenges and Actions

Restoration of the Moune wetlands has only just begun, and it is likely that we will be forced to constantly review our plans in response to a range of unpredictable political and economic factors. The issue of how to use the wetlands created by the earthquake and tsunami can only be decided by the local residents themselves. However, nature, including wetlands, offers much of value that is essential for human life. We believe it is science's role and responsibility to society to explain this value in a clear, straightforward manner. In Japan there is a tendency to classify things as "useful" or "not useful" according to the direct benefit they deliver to humans. This mindset means that, although people realize that environmental conservation is a positive and important thing, it is often difficult to gain support for it.

We believe that, rather than complaining, it is best to begin by accepting the situation as it is. We also believe that the first step toward restoration of the wetlands is identifying and communicating what the tidal flats offer that is of value, focusing on the *asari* clams that served as the impetus for restoration efforts. We hope that this first step will help to stimulate discussion about how Japan's coastal areas should be used.

.

Part III
Reconstruction from Radioactive Contamination

Chapter 8
Initiatives by the Soil Fertilization Team to Develop Agricultural Technologies for Paddy Fields with Radioactive Contamination

Itsuo Goto and Kaisei Inagaki

Abstract In April 2012, the paddy field located 20.8 km from the Fukushima Daiichi Nuclear Power Station was divided into six plots to apply zeolite and potassium chloride, and a cultivation test of rice was carried out. The concentration of radiocesium in the brown rice was highest in the control plot at 17 Bq/kg, but in plots where zeolite and potassium fertilizer were applied, the concentrations had decreased to 5–6 Bq/kg. The efficacy of zeolite efficacy in inhibiting the absorption of radiocesium by rice plants is considered to be mainly the result of its ability to adsorb ammonium and potassium ions, thereby inhibiting the leaching and outflow of nitrogen and potassium from the plow layer.

Keywords Radioactive contamination of farmlands • Radiocesium • Zeolite • Rice plant

8.1 Countermeasure for the Amelioration of Paddy Fields with Radioactive Contamination Jointly with Minamisoma Farmers

8.1.1 Radioactive Concentration of Paddy Field on Which a Test Planting of Rice Was Carried Out

We measured radioactive concentration in plow layer samples (at 15 cm) taken from a paddy field located in the city of Minamisoma, 20.8 km from the Fukushima Daiichi Nuclear Power Station, and the total radiocesium concentration was

I. Goto (✉) • K. Inagaki
Department of Applied Biology and Chemistry, Tokyo University of Agriculture,
1-1-1 Sakuragaoka, Setagaya-ku, Tokyo 156-8502, Japan
e-mail: igoto@nodai.ac.jp

© The Author(s) 2015
T. Monma et al. (eds.), *Agricultural and Forestry Reconstruction
After the Great East Japan Earthquake*, DOI 10.1007/978-4-431-55558-2_8

Fig. 8.1 Radioactively contaminated paddy field on which was carried out the rice planting test on Minamisoma City, Fukushima Prefecture (October 2011)

approximately 2,600 Bq/kg, comprising 1,137 Bq/kg of [134]Cs and 1,428 Bq/kg of [137]Cs (Fig. 8.1).

In April 2011, the Ministry of Agriculture, Forestry and Fisheries prohibited any rice planting for consumption purposes that year in paddy fields with a radiocesium concentration of 5,000 Bq/kg or more in the plow layer, or within a 30-km radius of the nuclear power station. The farmer was unable to plant this paddy field for the latter reason, and it turned into a field full of weeds.

As of October 2011, it was still unclear whether rice planting would be allowed in Minamisoma during 2012. However, we decided to take steps to inhibit the absorption of radiocesium by rice in this paddy field, in line with the farmer's wishes.

8.1.2 Application of Zeolite, Layer Mixing, and Soil Reversion

In December 2011, we applied zeolite and mixed the soil layers in the paddy field (Fig. 8.2). We divided 30 ares of the paddy field into three plots to which we applied either 10 or 20 t/ha of zeolite, or no zeolite (the control plot). The natural zeolite we

Fig. 8.2 Application of zeolite into paddy field (*left* photograph) and plow layer with mixing tillage (*right* photograph) (December 2011)

used for the test was produced by the Iizaka mine of Nitto Funka Kogyo, Co., Ltd. in Koori, Fukushima Prefecture. It had a cation-exchange capacity (CEC) of approximately 150 mEq/100 g with mordenite as its main component.

8.1.3   Test Planting of Rice

8.1.3.1   Application of Zeolite and Potassium Fertilizer to Inhibit Absorption of Radiocesium

In April 2012, these paddy fields were divided into six plots to apply zeolite and potassium chloride in different combinations of concentrations: three plots with a standard level of potassium chloride (equivalent to 50 kg/ha as K_2O) combined with either 0, 10, or 20 t/ha of zeolite, and three plots with a high level of potassium chloride (equivalent to $50 + 200 = 300$ kg/ha as K_2O) combined with either 0, 10, or 20 t/ha of zeolite.

In all test plots we applied a combination of legally registered fertilizers (50 kg/ha nitrogen, 40 kg/ha phosphate, and 50 kg/ha potassium), and for the high-potassium test plots, we added extra potassium chloride equivalent to 250 kg/ha as K_2O. On May 20, we planted rice of the Hitomebore variety in the paddy field.

It has already been established that the absorption of radiocesium can be inhibited by application of potassium fertilizer in the soil to promote the absorption of potassium by plants, because cesium exhibits the same behavior as potassium, which belongs to the same family of elements. Research to date indicates that radiocesium absorption by rice can be inhibited if the soil contains 25 mg/100 g of exchangeable potassium (K_2O) or more. Table 8.1 shows the levels of exchangeable potassium measured in plow layer samples taken while the rice was growing in June, July, and August. Exchangeable potassium amounted to 18 mg/100 g in the standard-potassium (50 kg/ha) no-zeolite (0 t/ha) plot in June, but it exceeded 25 mg/100 g in the other test plots. The reason why the standard potassium plots with zeolite also witnessed an increase in exchangeable potassium was because

Table 8.1 Temporal changes of exchangeable potassium (K_2O) in plow layer soil taken while the rice was growing in June, July, and August

Plots		Exchangeable K_2O (mg/100 g)		
Zeolite (t/ha)	K_2O (kg/ha)	June	July	August
0	50	18	11	18
0	300	26	22	16
10	50	28	20	15
10	300	40	26	17
20	50	43	23	18
20	300	43	37	25

Table 8.2 Radiocesium concentration in rice straw of the panicle formation stage and brown rice

Plots		Straw (Bq/kg)			Brown rice (Bq/kg)		
Zeolite (t/ha)	K_2O (kg/ha)	^{134}Cs	^{137}Cs	Total	^{134}Cs	^{137}Cs	Total
0	50	24.4	37.6	62.0	6.1	10.5	16.6
0	300	7.6	12.4	20.0	2.5	3.9	6.4
10	50	14.2	20.4	34.6	4.5	7.3	11.8
10	300	7.8	12.2	20.0	2.4	3.6	6.0
20	50	7.6	12.7	20.3	2.4	3.9	6.3
20	300	6.7	9.1	15.8	2.0	3.3	5.3

Measuring equipment: germanium detector
Detection limit
Straw: 10 Bq/kg (12-h measurement with U8 container)
Brown rice: 5 Bq/kg (80-min measurement with the marinade re-container of 2 l)

zeolite contains exchangeable potassium. Subsequently, as the rice grew, the amount of exchangeable potassium in all the test plots decreased, reaching about 15–25 mg/100 g in August.

8.1.3.2 Radioactive Concentration of Rice (Brown Rice and Straw)

Table 8.2 shows the concentration of radiocesium in the rice straw (stems and leaves of the panicle formation stage) sampling in July (Fig. 8.3), as well as the harvested brown rice, in September 2012. The concentration of radiocesium in the brown rice was highest in the standard-potassium, no-zeolite plot at 17 Bq/kg, but in plots where zeolite and potassium fertilizer were applied, the concentrations had decreased to 5–6 Bq/kg. The radiocesium detected in the rice straw sampled in July and August was three times the concentration found in the brown rice sampled at the same time. However, as with the brown rice, the concentrations of radiocesium in straw from the plots applied with both zeolite and potassium fertilizers were lower than in straw from the plots without zeolite. The detection limits were approximately 10 Bq/kg for the rice straw and approximately 5 Bq/kg for the brown rice,

Fig. 8.3 Rice of the panicle formation stage in paddy field of the rice planting test (July 2012)

and the quantification limits were about three times higher. The only test plot in which the concentrations of radiocesium exceeded the quantification limits was the standard-potassium, no-zeolite plot. Because the concentrations in all the other test plots were below these limits, it was impossible to compare the measurements quantitatively. However, it was clear that the application of a potassium fertilizer, or zeolite, or a combination of these materials, was effective in inhibiting the absorption of radiocesium. Nevertheless, zeolite efficacy in reducing radiocesium content when applied alone or together with a potassium fertilizer, and the mechanism by which it does so, were still not clearly known.

8.1.3.3 Investigating the Mechanism of Zeolite from the Yield and Taste Rating of Brown Rice

Table 8.3 shows the yield of brown rice together with taste appraisal values taken by a rice taste analyzer (Shizuoka Seiki PS-500). As shown in Fig. 8.4, the rice yield increased with the application of zeolite and potassium fertilizers. In addition, the protein content increased as the yield increased, and a high level of correlation is observed between the two, as shown in Fig. 8.5. Upon analyzing the factors causing the increase in protein content, we found no increase in protein content resulting

Table 8.3 Yield of brown rice and taste appraisal values taken by rice taste analyzer[a]

Plots		Yield	Taste appraisal values				
Zeolite	K₂O		Moisture	Protein	Amylose	Fatty acid	
t/ha	kg/ha	t/ha	%	%	%	mg/100 g	Score
0	50	4.49	15.8	6.0	19.8	13.3	78.7
0	300	4.93	15.5	6.1	19.9	13.7	76.7
10	50	5.30	16.0	6.4	20.2	14.7	73.7
10	300	6.09	16.3	6.4	20.0	15	73
20	50	5.53	15.9	6.2	19.9	14	76.7
20	300	6.76	16.3	6.5	19.8	15.3	72.3

[a]Rice taste analyzer (Shizuoka Seiki PS-500)

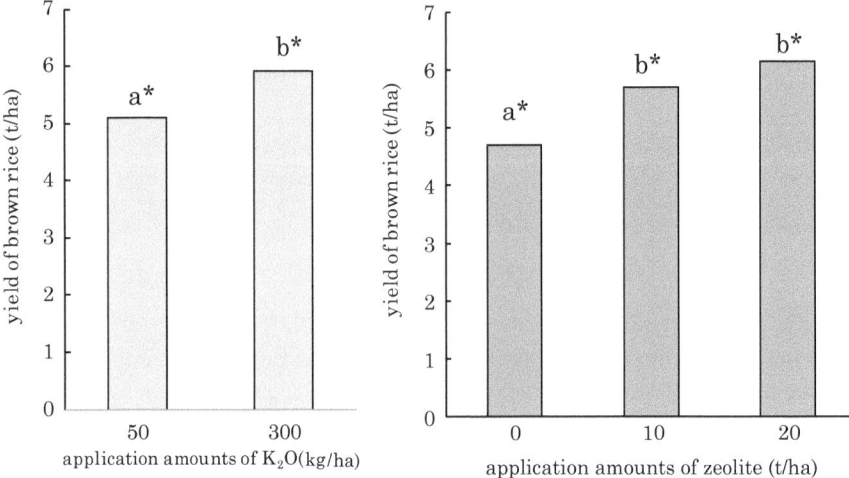

Fig. 8.4 Effects of zeolite or potassium application on yield of brown rice. *Different letters indicate significant differences (Tukey's q test, $p<0.25$)

from the application of potassium (as shown in Fig. 8.6), but we found that the protein content tended to rise with the application of zeolite. It is well known that the taste appraisal values of brown rice drops if the protein content increases and a significant correlation between the taste appraisal values and protein content was observed in this test as well. When the factors behind the drop in taste rating were analyzed, the results showed that the rating tended to drop with the application of zeolite, as shown in Fig. 8.7.

These results indicated that the application of zeolite promoted the absorption of nitrogen fertilizer in the rice plants, bringing about an increase in the protein content and yield of the brown rice. It was considered that the increased efficiency of nitrogen fertilizer application resulting from zeolite could be related to the inhibition of radiocesium absorption.

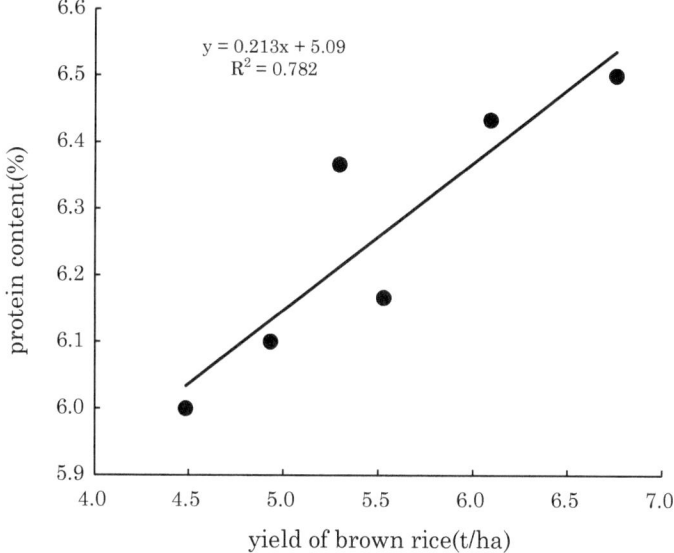

Fig. 8.5 Relationship between yield and protein content of brown rice

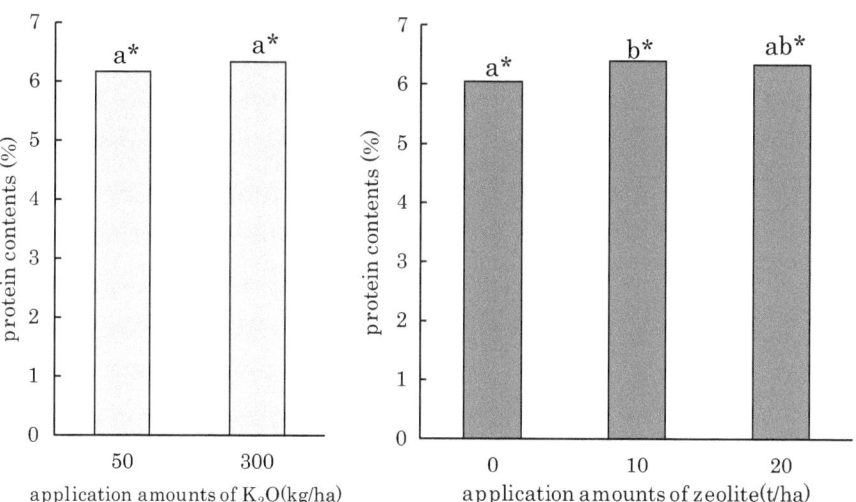

Fig. 8.6 Effect of zeolite or potassium application on protein content of brown rice. *Different letters indicate significant differences (Tukey's q test, $p < 0.25$)

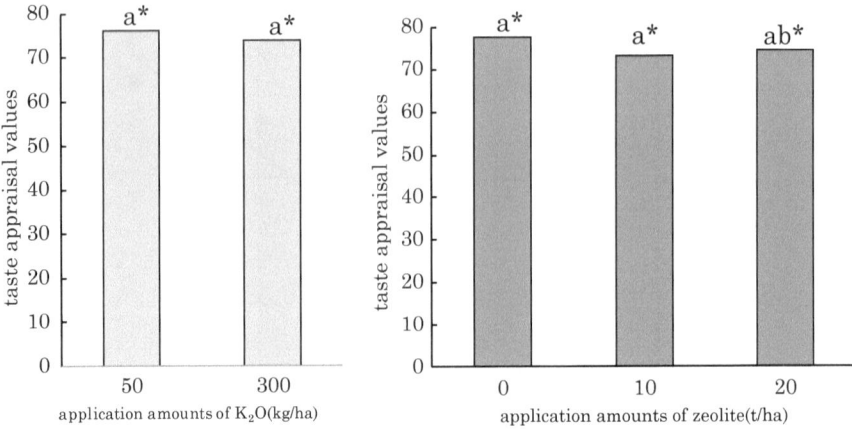

Fig. 8.7 Effects of zeolite or potassium application on taste appraisal values of brown rice. *Different letters indicate significant differences (Tukey's q test, $p < 0.25$)

8.2 Countermeasure for the Amelioration of Paddy Fields with Radioactive Contamination in Date City

8.2.1 Test Planting of Rice in Date

In the fall of 2011, a large quantity of rice from the mountain paddy fields in the Abukuma Highlands in the city of Date, Fukushima Prefecture, was found to exceed the government's provisional radiation limit of 500 Bq/kg. As a result, rice planting for consumption purposes was prohibited in six districts within the city in 2012.

In a tie-up with JA Datemirai, a local agricultural cooperative, we carried out rice planting tests to investigate the application effects of zeolite and potassium fertilizers, or amelioration of soil acidity, on the absorption of radiocesium to rice plants.

8.2.2 Method of Rice Planting Test

As shown in Table 8.4, two adjacent plots were set up as test plots in each of three locations (A, B, and C in Fig. 8.8) in the Kakeda district of Ryozen, a town in Date. In Kakeda Paddy A, which had a radiocesium concentration of 7,600 Bq/kg in the plow layer in April 2011, we studied the application effect of zeolite alone. In Kakeda Paddy B, which had a radiocesium concentration of 8,700–10,360 Bq/kg, we compared different combinations of zeolite and two different types of potassium fertilizer (potassium chloride and potassium silicate). In Kakeda Paddy C, which had a radiocesium concentration of 4,900 Bq/kg, we observed the application effect of converter slag for amelioration of soil acidity. In the middle of May we applied

Table 8.4 Results of test planting of rice in three paddy fields

Paddy field	Application materials (t/ha)				Soil analysis				Conc. of $^{134}Cs+^{137}Cs$		Brown rice		
	Zeolite	Potassium silicate	KCl	Converter slag	$^{134}Cs+^{137}Cs$ Bq/kg	pH (H$_2$O)	K$_2$O mg/100 g	CEC meq/100 g	Straw Bq/kg	Brown rice Bq/kg	Yied t/ha	Protein %	Taste appraisal values
A1	10	2	–	2	7,569	6.3	27.9	14.2	40.4	4.0	4.59	5.6	83
A2	–	2	–	2	7,569	6.3	22.3	13.9	37.6	5.4	3.44	5.4	85
B1	10	2	–	–	8,734	6.2	57.6	37.6	35.7	2.4	4.48	5.4	85
B2	2	–	0.67	–	10,190	6.1	34.9	34.8	37.9	3.8	4.57	5.3	86
B3	2	2	–	–	10,360	6.1	18.0	33.6	73.9	11.2	4.15	5.3	86
C1	–	2	–	10	4,899	7.5	8.1	17.8	104	29.3	4.82	6.2	75
C2	–	2	–	–	4,899	6.0	7.9	13.8	182	58.2	4.14	5.7	81

Fig. 8.8 Paddy fields for rice planting test in Date City

the materials to treat the soil and replanted rice of the Koshihikari variety into the paddy fields, harvesting the rice on September.

8.2.3   Test Results

As shown in Table 8.4, we observed no changes in radiocesium concentration from the application of 10 t/ha of zeolite in paddy field A. However, the protein content and rice yield increased while the taste rating dropped. In addition, the water-soluble potassium content of the exchangeable potassium in the soil after applying fertilizer and after harvesting was lower in the plot where zeolite was applied than in the plot without zeolite.

In paddy field B, the protein content and the taste rating in the brown rice were almost the same between the three test plots, but the rice yield was higher in the plot where 10 t/ha rather than 2 t/ha of zeolite was applied and the plot where potassium chloride was applied. In addition, the radiocesium concentration in the stems and leaves, as well as in the brown rice, dropped in both these plots. Application of zeolite, or potassium chloride as a potassium fertilizer, was effective for inhibition of the radiocesium to rice in this paddy field.

In paddy field C, the application of 10 t/ha converter slag resulted in radiocesium concentration in the stems and leaves, as well as in the brown rice, decreasing by approximately half. On the other hand, the protein content in the rice increased, while the yield also increased by 16 %. The taste rating dropped in line with these changes.

As shown here, no effect on radiocesium levels from application of zeolite alone was evident in paddy field A. In paddy field B, the application of zeolite and potassium chloride resulted in reduction of the radiocesium concentration in the stems and leaves, as well as in the brown rice, but the mechanism was not evident. Nevertheless, an increase in the protein content and yield was observed with the application of zeolite in both paddy fields A and B. This phenomenon is the same as the test result obtained in Minamisoma, and its mechanism can probably be explained by the increase in efficiency of nitrogen fertilizer resulting from the well-known property of zeolite for specific adsorption of ammonium ion.

In paddy field C, the pH of the soil rose from 6.0 to 7.5 with the application of converter slag. It is known that when the soil pH increases, mineralization of nitrogen is accelerated by the alkali effect. The observed increase in the rice yield could also be caused by this alkali effect. In addition, the reduction in the radiocesium concentration in the brown rice is considered to have been caused by a gradual increase in the ammonium ion concentration in the soil, leading to an inhibition in the absorption of radiocesium by the rice as a result of increased competition with the cesium ions.

8.3 Is Application of Zeolite to Paddy Fields Effective as a Soil Amendment for Inhibition of Radiocesium Absorption in Rice?

8.3.1 Cesium-Specific Adsorption Characteristics by Zeolite

We added 1 g zeolite to 100 ml cesium chloride solutions containing 1, 50, or 100 mg/l cesium (^{133}Cs) in 100-ml polyethylene containers and left the solutions to stand for 24 h. Subsequently, the solutions were filtered and the ^{133}Cs in the filtrate measured using a quadrupole ICP mass spectrometer. Similar experiments were also carried out for sodium ions and potassium ions of the same concentration for comparison with the cesium. The zeolite used was a natural zeolite with a particle size of 1–2 mm that contained mainly clinoptilolite produced by the Itaya mine in Yonezawa, Yamagata Prefecture. The cesium adsorption characteristics of the zeolite were stronger than potassium, which was a homologous element (Table 8.5). Within 24 h, just 1 g zeolite adsorbed almost all the cesium in 100 ml cesium solution with a concentration of 100 mg/l.

Next, 1 g zeolite was added to 100 ml cesium chloride solution containing 100 mg/l ^{133}Cs in a 100-ml polyethylene container and then shaken in a shaker for 5

Table 8.5 Difference in alkali metals ion adsorption ratio of the zeolite

Alkali ion	Sodium		Potassium		Cesium	
Concentration	Residual conc.	Adsorption ratio	Residual conc.	Adsorption ratio	Residual conc.	Adsorption ratio
mg/l	mg/l	%	mg/l	%	mg/l	%
1	0.54	45.8	0.25	74.9	0.00	99.7
50	38.7	22.5	1.72	96.6	0.02	99.9
100	88.1	11.9	3.82	96.2	0.06	99.9

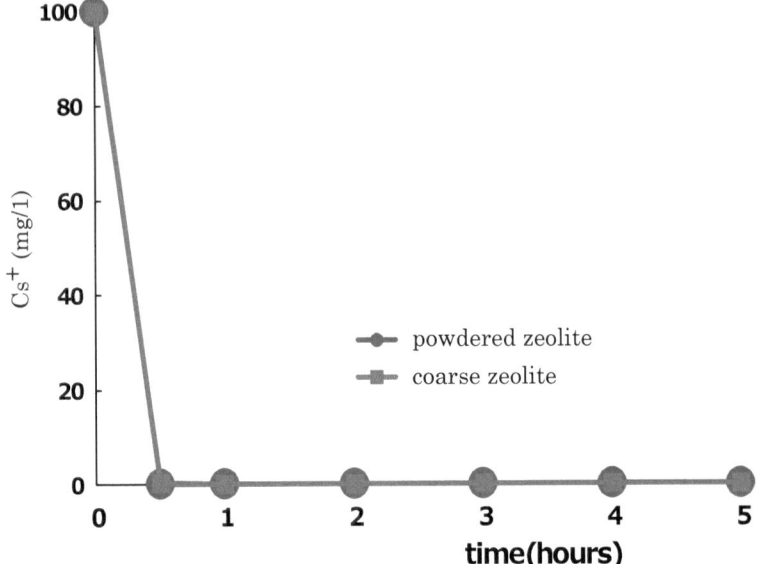

Fig. 8.9 Cesium ion adsorption speed of zeolite

h. After the initial 30 min, the container was taken out of the shaker every hour to be filtered, and then the concentration of ^{133}Cs in the filtrate was measured. Two types of zeolite were used. One was the same coarse zeolite used in the first experiment and the other was the coarse zeolite ground into powder form. As shown in Fig. 8.9, almost all the ^{133}Cs was adsorbed after just 30 min irrespective of the difference in the zeolite particle size. This result indicated that the zeolite more specifically than potassium adsorbs radiocesium in the radioactively contaminated soil.

Our next aim was to demonstrate the behavior of cesium after it has been adsorbed by zeolite. We obtained three types of natural zeolite produced in different locations in Japan (clinoptilolite from Yamagata Prefecture, mordenite from Fukushima Prefecture, and clinoptilolite from Hokkaido). These three types of zeolite and 0.5 g each of 22 soils collected from various regions were added individually to 50-ml samples of an aqueous solution containing 1 mg/l ^{133}Cs. The 25 samples were then shaken for 1 h before the cesium adsorption ratio was measured. Next, 50 ml 1 M/l ammonium acetate with a pH of 7 was added to these solutions containing zeolite or soil. After shaking for 1 h, the concentration of ^{133}Cs in the filtrate was measured, and then the proportion of cesium that was exchanged and released by the ammonium ions (the release ratio) was calculated. The adsorption ratio of all three types of zeolite was 99.8 % or more (Fig. 8.10). The adsorption ratio for soil was also high, at 94–98 %. However, compared to a release ratio of 26–33 % for zeolite, the release ratio for soils was higher at 48–99 %. In other words, although soils may have an adsorption capability on a par with that of zeolite, zeolite is considered to release less cesium than soils.

As potassium and calcium ions coexist in large quantities in farmland soils, we compared the cesium adsorption properties of zeolite and soil in coexistence with potassium ions. High-humic Andosol sampled from uncultivated land in Kanuma, Tochigi Prefecture; chernozem sampled from wheat fields in Ukraine; and Kanuma pumice were used as the test soils.

We added 0.5 g zeolite and the aforementioned soils to 50 ml each of two different solutions: a solution containing 1 mg/l ^{133}Cs only, and an solution containing a mixture of 1 mg/l ^{133}Cs and 100 mg/l K. After shaking for 1 h, the solutions were filtered and the ^{133}Cs measured. As shown in Fig. 8.11, the soils showed a decrease in the cesium adsorption ratio as a result of the coexistence of potassium ions, whereas the zeolite showed no change at all. In other words, zeolite can easily adsorb cesium even in soils with a high concentration of potassium ions, such as cultivated land.

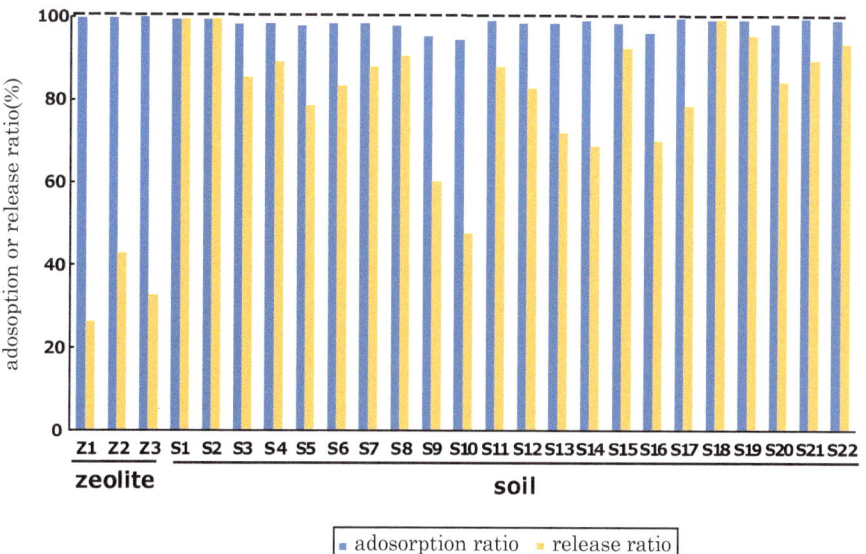

Fig. 8.10 Cesium ion adsorption and release ratio of zeolite and soils

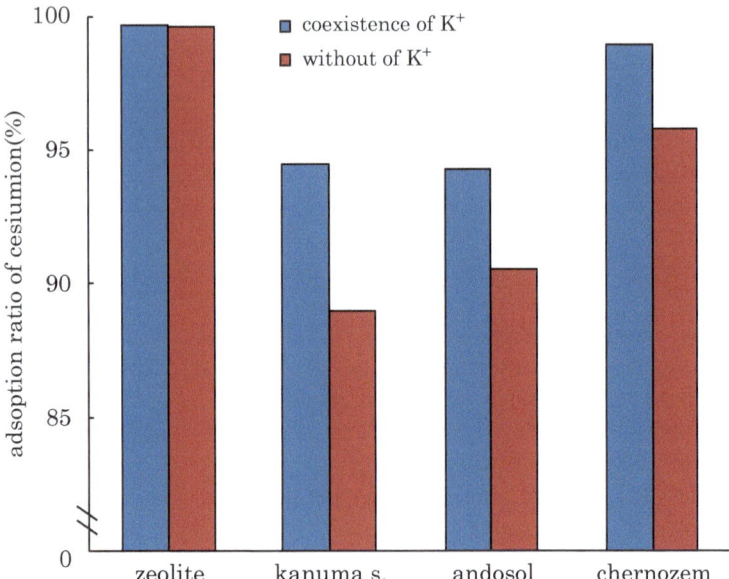

Fig. 8.11 Effects of coexistence of potassium ion in cesium ion adsorption properties of zeolite and soil

The results of this experiment indicate that when zeolite is applied to soil contaminated with radiocesium, the zeolite adsorbs water-soluble cesium ion, which is easily taken up by a plant, and cesium ions that are weakly exchanged and adsorbed by negative charges of humus or clay minerals. If nitrogen and potassium fertilizers are applied before rice planting, ammonium ions and potassium ions are adsorbed by the zeolite. Although some cesium ions may be released, the release of cesium into the soil solution is deemed to be less than in soil without zeolite.

8.3.2 Inhibition Mechanism of Radiocesium Absorption to Rice by Application of Zeolite

The effectiveness of potassium fertilizers in inhibiting the absorption of radiocesium by crops has been reported on many occasions since the atom and hydrogen bomb tests on Bikini Atoll, the Chernobyl nuclear disaster, and the Fukushima Daiichi Nuclear Power Station accident. Ammonium ions, on the other hand, which are almost the same size and behave the same way as potassium ions in soil, are often considered as having no effect in inhibiting the absorption of radiocesium or, conversely, may even be considered as accelerating its absorption. The results of the foregoing tests conducted by the authors showed an increase in the protein content and yield of brown rice from the application of zeolite in the paddy fields and an accompanying reduction in the radiocesium concentration. Moreover, the results of laboratory tests on the behavior of potassium ions in soil where potassium chloride has been applied suggest that the mechanism of the zeolite effect is as follows.

Being an alkali metal, cesium behaves as a univalent cation in soil. As a result, it exists as exchangeable cesium adsorbed by the negative charges in soil colloids and water-soluble cesium dissolved in the soil solution. Compared to the cations that exist in large quantities in the soil, such as calcium, magnesium, and potassium, the specific adsorption strength (exchange invasion power) of cesium in the soil colloids is higher, and thus the proportion of water-soluble cesium is extremely low. When zeolite is applied to paddy fields contaminated with radiocesium, water-soluble radiocesium ions that have not been adsorbed by the clay minerals are adsorbed by the zeolite. As a result, the concentration of water-soluble radiocesium drops, and thus radiocesium ions that have been adsorbed by pH-dependent charges in the humus or clay minerals separate and are adsorbed by the zeolite. Consequently, a plot where zeolite is applied has a lower concentration of radiocesium ions in the soil solution than a plot without zeolite. If ammonium and potassium ions are supplied as fertilizers, both ions behave in the same way as radiocesium ions and thus the concentration of ammonium and potassium ions in the soil solution drops. Because radiocesium ions are adsorbed more easily by the zeolite than ammonium ions and potassium ions, the ratio of $NH_4^+ + K^+$ to Cs^+ in the soil solution is higher in plots where zeolite has been applied. When the root of the rice plant extends into soil having such an ion composition, the plant will preferentially absorb ammonium and potassium ions over radiocesium ions and thus the absorption of radiocesium can be inhibited. If the concentration of both ammonium and potassium ions in the soil solution drops after being absorbed by the rice plant, those ions that have been adsorbed in the zeolite and soil will separate to maintain the ion concentration ratio at a certain level. Therefore, the absorption of radiocesium ions by the rice plant is inhibited until no more ammonium and potassium ions are left in the soil.

Paddy fields generally have a water requirement in depth of about 20 mm/day. Although some of the water may evaporate from the surface of the paddy fields, a proportion will permeate downward. As a result, some of the water-soluble ammonium and potassium ions applied as fertilizers will leach into the sublayers. In addition, water can flow from the paddy fields, particularly during heavy rains, and these ions may also flow out as a result. However, as the laboratory test results show, if zeolite is applied, it will adsorb both the ammonium and potassium ions, resulting in fewer water-soluble ions that can leach or flow out, thereby maintaining the inhibition of radiocesium absorption by the rice plants. In addition, application of zeolite is thought to have increased the efficiency of the nitrogen and potassium fertilizers, resulting in an increase in the protein content and yield of the brown rice.

In other words, zeolite is able to specifically adsorb cesium ions, and this specificity exceeds that for ammonium and potassium ions. However, the efficacy of zeolite in inhibiting the absorption of radiocesium by rice plants is considered to be mainly because of its ability to adsorb ammonium and potassium ions, thereby inhibiting the leaching and outflow of nitrogen and potassium from the plow layer.

When trying to inhibit the absorption of radiocesium in paddy fields, therefore, instead of applying a large quantity of potassium fertilizers from the start, zeolite should be applied before determining the amount of potassium to use.

Zeolite can act to adsorb ammonium and potassium ions in paddy fields when used to inhibit the absorption of radiocesium in crops. In upland fields where almost no ammonium ions exist because of the formation of nitrate ion, zeolite functions by adsorption of potassium ions. The optimal amount to apply is 5–10 t/ha, or alternatively, approximately 0.2–0.3 t/ha can be applied continuously for several years. Additional applications are not needed as zeolite remains in the soil for the long term as a stable mineral.

Natural zeolite sold commercially varies in terms of production area, main mineral component (mordenite, clinoptilolite), particle size, and other factors. However, in Japan there is no need to rely on imports and artificial zeolite as the country produces a large quantity of world-class natural zeolite (with a CEC of approximately 150 mEq/100 g), mainly in the Tohoku region, including Fukushima, Yamagata, Miyagi, and Akita prefectures. So long as the CEC is high, there is no difference in efficacy according to zeolite particle size or main mineral component. That said, powdered products may be cheap, but they scatter easily during application. In contrast, coarse products with a particle size of about 1–2 mm are easier to apply but are more expensive. The most efficient way to apply zeolite, therefore, is to use a lime sower to scatter a low-cost powdered product that has not been overdried, but contains approximately 20 % water.

Chapter 9
The Potential for Producing Rice for Feed and Whole-Crop Rice Silage in Radiation-Contaminated Areas

Seiji Nobuoka

Abstract Neither whole rice plants for silage nor unhulled rice grains can be used as livestock feed if their radiation concentration exceeds the threshold limit of 100 Bq/kg. However, the results outlined here confirm that even in paddies with a comparatively high radioactive cesium soil concentration of 2,600 Bq/kg, this threshold can be met by plowing and applying zeolite as a decontamination measure. This finding is positive news for areas that have suffered radioactive contamination. The study also found the level of radiation in unhulled grains of fodder rice to be below the 100 Bq/kg limit, meaning that the rice was usable as livestock feed.

Keywords Cesium • Rice • Feed • Silage • Livestock • Zeolite • Potash

9.1 Introduction: Research Objectives

On August 9, 2013, the Japanese Ministry of Agriculture, Forestry and Fisheries released a map showing concentrations of radionuclide in farmland soil. According to this map, soil radiation concentration exceeds 5,000 Bq/kg of radioactive cesium-134 and -137 combined in an estimated 7,500 ha of agricultural land in Fukushima Prefecture, requiring decontamination by physical means. Additionally, an estimated 53,822 ha of agricultural land in Fukushima Prefecture is contaminated with a concentration of 1,000–5,000 Bq/kg. Rice paddies account for three quarters of this area. Since 2011, in line with the policy on rice planting set by the Japanese government's Nuclear Emergency Response Headquarters, cultivation of rice for consumption has been prohibited in paddies where radioactive contamination exceeds 5,000 Bq/kg. This policy is based on an estimated radiation transfer

S. Nobuoka (✉)
Laboratory of Livestock Farming Management, Department of Animal Science, Faculty of Agriculture, Tokyo University of Agriculture,
Funako 1737, Atugi, Kanagawa 243-0034, Japan
e-mail: s3nobuok@nodai.ac.jp

© The Author(s) 2015
T. Monma et al. (eds.), *Agricultural and Forestry Reconstruction After the Great East Japan Earthquake*, DOI 10.1007/978-4-431-55558-2_9

factor from soil to unpolished rice of 0.1, and the planting prohibition remained in place as of 2013.

In 2012 the same prohibition was also placed on some paddies with a soil cesium concentration below 5,000 Bq/kg because the level of radioactive cesium permissible in unpolished rice for human consumption had been lowered from 500 to 100 Bq/kg. In 2013 paddy land has been divided into three broad categories based on the level of soil contamination: (1) planting prohibited (other than for experimental/nonconsumption purposes), (2) test planting permitted (in preparation for resumption of cultivation), and (3) cultivation and shipment permitted (under fully controlled conditions). With the establishment of these categories, a path toward agricultural recovery has begun to emerge.

The Laboratory of Livestock Farming Management is part of the Department of Animal Science at Tokyo University of Agriculture. Since 2011 the laboratory has been participating in the university's East Japan Assistance Project, working to develop decontamination measures with other laboratories such as the Department of Applied Biology and Chemistry's Laboratory of Agricultural Production and Environmental Chemistry.

The associated research focuses on rice grown for livestock feed in paddies with a radiation concentration of 5,000 Bq/kg or less. Its aim is to develop a method that uses potassium fertilizer and zeolite material to prevent rice for feed from absorbing radioactive cesium, and to confirm its safety for use as feed by giving the harvested product (rice for feed or whole-crop rice silage) to livestock for consumption.

9.2 Tests to Prevent Cesium Absorption in Livestock Feed Rice

We believe that, in addition to physical decontamination measures, the development of methods to prevent transfer of radioactive cesium from soil to crops is an effective means of assisting the areas affected by the nuclear accident. However, the majority of research into rice paddy decontamination methods has focused on rice for human consumption, and very little testing and research has been conducted thus far in relationship to crops such as rice for livestock feed.

Recent years have witnessed rapid growth in the Japanese land area used for cultivation of domestically grown livestock feed in the form of rice grains for fodder and whole rice plants for silage. Nationwide, 33,726 ha of fodder rice and 31,157 ha of rice plants for silage were produced in 2014, amounting to a total of 64,883 ha. In Fukushima Prefecture, meanwhile, 893 ha of fodder rice and 762 ha of rice plants for silage were produced in 2014. Of course, rice for livestock feed and other crops can be grown in paddies with a soil radiation level of 5,000 Bq/kg or lower in the same way as rice for human consumption. It is therefore essential to develop methods to prevent cesium from being absorbed into the stalks, leaves, and grains used for livestock feed to ensure the safety of the products derived from animals to which it is fed. To this end, we began conducting tests to prevent cesium absorption in rice paddies in the Haramachi district of Minamisoma City in Fukushima Prefecture. The following is an overview of the test site.

	Potassium 5kg/10a				Potassium 30kg/10a	
Zeolite 2 t /10a	Momiroman	Takanari	hokuriku 193	Hitomebore	Hitomebore	Koshihikari
Zeolite 1 t /10a	Momiroman	Takanari	hokuriku 193	Hitomebore	Hitomebore	Koshihikari
Zeolite 0 t /10a	Momiroman	Takanari	hokuriku 193	Hitomebore	Hitomebore	Koshihikari

Fig. 9.1 Allocation of test plots and rice varieties

1. Test site

    Rice paddies covering 30 ares in the Haramachi district of Minamisoma City, Fukushima Prefecture (20.8 km from the Fukushima Daiichi nuclear power station)

2. Allocation of test plots

    Zeolite was differentiated by volume into control plots (with no zeolite applied) and test plots (with either 1 t/10 ares or 2 t/10 ares of zeolite applied). Potash fertilizer (potassium chloride) was applied in the zeolite plots, differentiating by volume into standard-potassium plots (5 kg/10 ares) and high-potassium plots (30 kg/10 ares).

3. Rice varieties tested

    We tested Momiroman, Takanari, and Hokuriku 193 as livestock feed varieties, and Hitomebore and Koshihikari as varieties for human consumption (Fig. 9.1).

4. We conducted growth studies and sampled rice plants and soil at each stage of growth.

5. We measured the concentration of radioactive cesium in sampled rice plants and soil using a germanium semiconductor detector.

9.3 Results of Tests to Prevent Cesium Absorption in Rice for Livestock Feed

The following are results of analysis of radioactive cesium concentrations (combined totals of cesium-134 and -137) in paddy soil at the test site in the Haramachi district of Minamisoma City, Fukushima Prefecture.

9.3.1 Radioactive Cesium Concentrations Measured in Test Paddies

The initial concentration of radioactive cesium at the test site (before plowing) was 2,600 Bq/kg. After the paddies were plowed to reduce radioactive cesium concentration and decontaminate the soil, then replowed with potash fertilizer with or without zeolite, radioactive cesium concentration in the paddy soil fell by half, to approximately 1,400 Bq/kg.

9.3.2 Radioactive Cesium Concentrations at Various Soil Depths

The concentration of radioactive cesium at various soil depths was measured while the rice plants were growing. In soil samples taken on July 19, 2012, concentrations ranged from 1,008 to 1,586 Bq/kg at 0–5 cm, from 798 to 1,641 Bq/kg at 5–10 cm, and from 212 to 1,113 Bq/kg at 10–15 cm. The concentration in the soil deeper than 15–20 cm was far lower, less than 200 Bq/kg (Fig. 9.2).

In the next soil samples taken on September 12, 2012, concentrations ranged from 1,183 to 1,696 Bq/kg at 0–5 cm, 1,102 to 1,782 Bq/kg at 5–10 cm, and 449 to 1,398 Bq/kg at 10–15 cm. Although levels in the deeper soil were lower, measuring

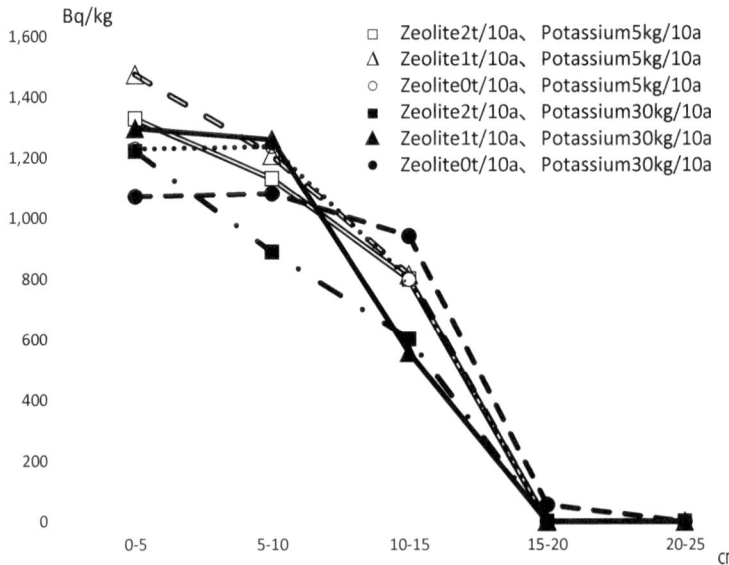

Fig. 9.2 Concentration of radioactive cesium at various soil depths (Samples taken July 19, 2012)

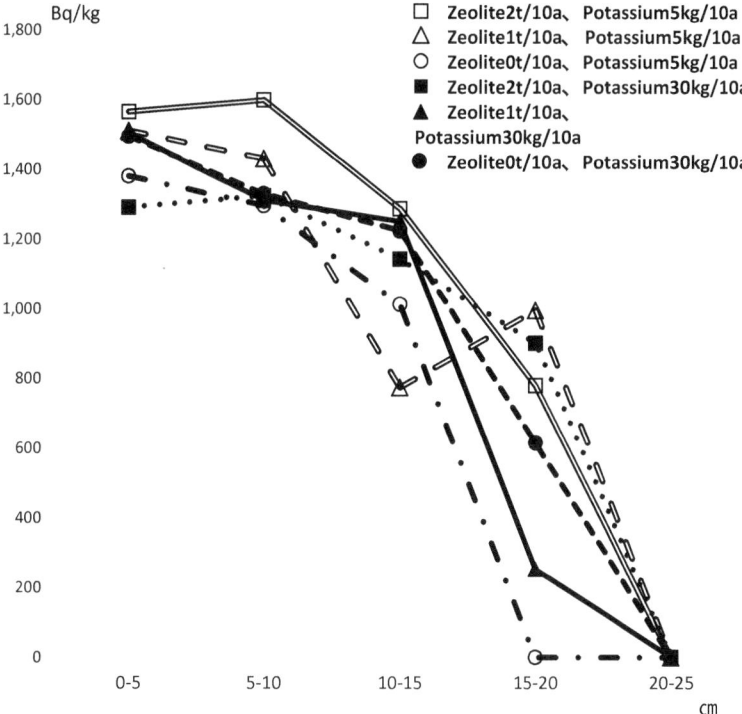

Fig. 9.3 Concentration of radioactive cesium at various soil depths (Samples taken September 12, 2012)

27 to 843 Bq/kg at 15–20 cm, and 12 to 155 Bq/kg at 20–25 cm, radiation concentration in the subsoil had gradually increased compared to the July samples (Fig. 9.3).

The foregoing results therefore show that radioactivity is moving to the subsoil, albeit gradually. Although the actual level of the radiation was not particularly high, the situation needs to be monitored in future as the roots of rice plants extend to the subsoil.

9.3.3   Radioactive Cesium Concentration in Whole Rice Plants for Silage

The tests conducted in Minamisoma City, Fukushima Prefecture, to study prevention of radioactive cesium absorption in rice for livestock feed, mainly focused on the application of zeolite. Our findings showed that only two varieties of rice plants for silage had a radioactive cesium concentration exceeding 100 Bq/kg at the time of harvest (September 12, 2012): Momiroman [in a no-zeolite, standard-potassium (5 kg/10 ares) plot], and Hokuriku 193 (also in a no-zeolite, standard-potassium plot). None of the samples from the plots applied with a combination of both zeolite

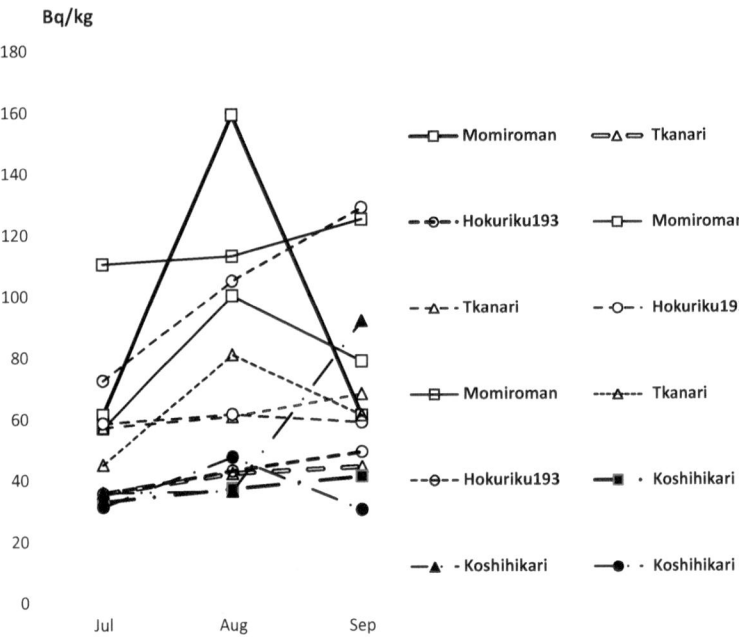

Fig 9.4 Radioactive cesium concentration in rice plants for silage by variety, stage of growth, and amount of zeolite/potash fertilizer applied

and potash fertilizer exceeded the 100 Bq/kg government safety limit for use as livestock feed (Fig. 9.4).

The average cesium concentration in rice plants for silage at the time of harvest in September was 51.8 Bq/kg in the 2-t zeolite plot, 68.9 Bq/kg in the 1-t zeolite plot, and 105.3 Bq/kg in the 0-t zeolite plot. Compared with the 0-t zeolite plot, absorption was reduced to 49 % in the 2-t zeolite plot and 65 % in the 1-t zeolite plot.

Variation among rice varieties was observed, with cesium concentrations in the 2-t zeolite plot of 61.4 Bq/kg in Momiroman, 49.6 Bq/kg in Hokuriku 193, and 44.6 Bq/kg in Takanari. Although definite conclusions cannot be drawn because of the small sample size, it is believed that Momiroman absorbs cesium more readily than other varieties such as Takanari and Hokuriku 193.

The radioactive cesium concentration was 41.5 Bq/kg in the unhulled grains of Koshihikari rice for human consumption grown as a control (in a plot containing 2 t/10 ares of zeolite and 30 kg/10 ares of potash). This finding suggests that the rice varieties used for livestock feed have a slightly higher tendency to absorb cesium than the varieties grown for human consumption.

Considering cesium concentration in the rice plants at various stages of growth, the July 19 level in the 2-t zeolite plot of Momiroman was 61.6 Bq/kg, whereas the concentration in the 0-t plot was approximately double this at 110.8 Bq/kg. This test result confirmed the effectiveness of zeolite in reducing cesium absorption. Additionally, the August and September analysis results showed a trend toward higher levels of radioactive cesium in plants compared to the July 19 analysis data.

However, the analysis data from September 12 returned scattered results, with cesium concentration declining in some plants compared to August and September while conversely increasing in others.

Whole rice plants for silage are harvested and bale-rolled in their entirety, including stalks and leaves, during the milk-ripe stage before the grains have hardened. It is therefore important that radioactive cesium concentration in August and September does not exceed the 100 Bq/kg safety threshold for livestock feed. However, there was an issue in this regard, as the concentration in the livestock feed rice varieties Momiroman and Hokuriku 193 exceeded 100 Bq/kg when zeolite was not administered. When cultivating rice for use as livestock feed, it is therefore necessary to administer large quantities of zeolite and potash fertilizer as a means of reducing radioactive cesium absorption.

9.3.4   Radioactive Cesium Concentration in Unhulled Rice Grains for Fodder

Concentrations of radioactive cesium in unhulled rice grains for use as fodder are outlined next. Radioactive cesium concentration in unhulled Takanari rice was 45.4 Bq/kg in the 2-t zeolite plot, 65.4 Bq/kg in the 1-t zeolite plot, and 47.5 Bq/kg in the 0-t zeolite plot (Fig. 9.5).

Cesium concentration in unhulled grains of Takanari fodder rice was below 100 Bq/kg in both test plots as well as the control plot. However, these results did not provide enough evidence to draw a clear conclusion that zeolite was effective in reducing absorption of radioactive cesium. We are currently continuing on-site tests using a combination of zeolite and potash fertilizer for the 2013 crop.

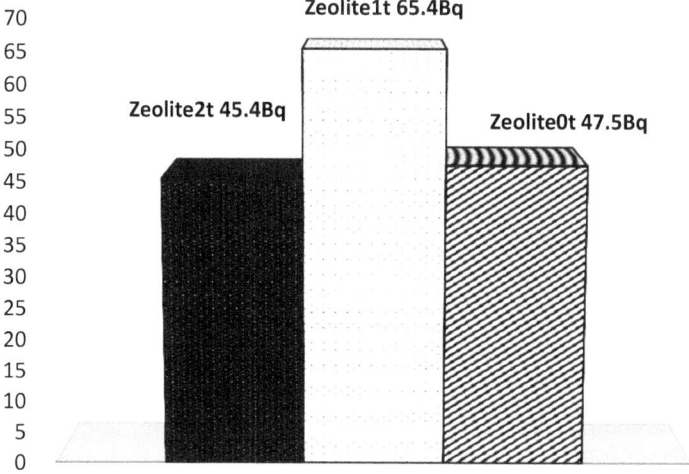

Fig. 9.5 Concentration of radioactive cesium in unhulled Takanari rice grains by amount of zeolite applied

9.4 Summary

Neither whole rice plants for silage nor unhulled rice grains can be used as livestock feed if their radiation concentration exceeds the threshold limit of 100 Bq/kg. However, the results outlined here confirm that even in paddies with a comparatively high radioactive cesium soil concentration of 2,600 Bq/kg, this threshold can be met by plowing and applying zeolite as a decontamination measure. This finding is positive news for areas that have suffered radioactive contamination. The study also found the level of radiation in unhulled grains of fodder rice to be below the 100 Bq/kg limit, meaning that the rice was usable as livestock feed.

Besides testing the efficacy of zeolite efficacy, we also considered the economic feasibility of zeolite application as a means of reducing cesium absorption. In addition to the labor required to apply zeolite, the material itself costs around 75 yen per kilogram. Application of 2 tons per 10 ares would therefore incur a cost of 150,000 yen per 10 ares for the material alone. Alternatively, application of 1 ton per 10 ares would cost 75,000 yen, and this amount would be eligible for compensation as decontamination costs. Zeolite is, moreover, straightforward to use: it is easily applied with a broadcast spreader and a single application fixes cesium, making application every year unnecessary. From an economic and practical point of view, the findings of this study can therefore be put into effect immediately.

With the 2013 crop, the planting prohibitions have been relaxed. Cultivation is possible even in contaminated areas, on condition that the land is decontaminated

Fig. 9.6 Italian ryegrass silage (34.8 Bq/kg)

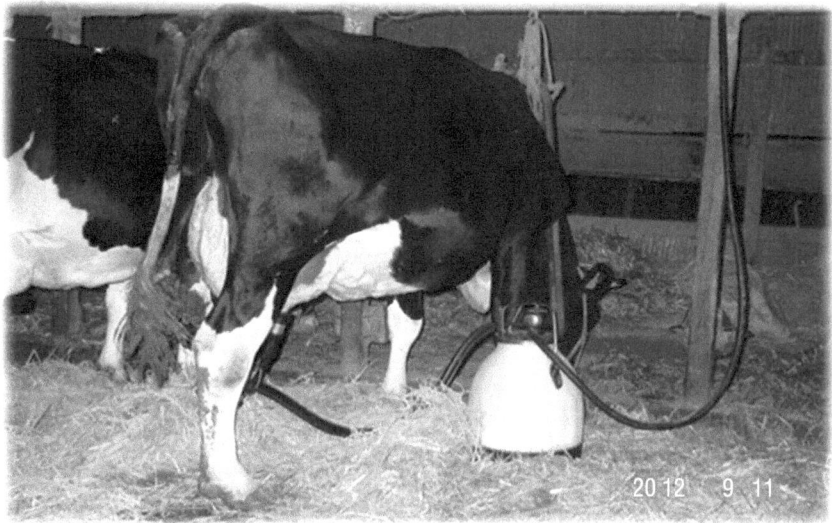

Fig. 9.7 Feeding test for dairy cows

by such means as deep plowing or reversing the soil and spreading potash material, whereas each bag of rice must be radiation tested before shipment. Zones have been designated where test planting is permitted in preparation for resumption of cultivation, and planting has recommenced in selected areas. The site used for this study in the Haramachi district of Minamisoma City was designated as a test planting zone in 2013, so there will be a rice harvest this year.

We are conducting our Haramachi study jointly with two local farmers, one of whom planted 8 ha of rice in 2013, leading the way toward the revival of agriculture in the district. After seeing the example set by this farmer and the tests conducted by Tokyo University of Agriculture, another farmer in the area has expressed a desire to begin growing rice for livestock feed in 2014. Thus, our tests are gradually helping to provide a foothold for getting the district's agricultural revival under way.

At the same time, actual numeric data are needed to verify the extent to which radiation transfers to livestock products even if animals are fed rice plants with a cesium concentration lower than the limit of 100 Bq/kg. In a 2012 study targeting dairy farmers in the Nasu area of Tochigi Prefecture, dairy cows were fed Italian ryegrass silage with a radioactive cesium concentration of 34.8 Bq/kg over a 6-month period. Subsequently, the cesium concentration in the raw milk was measured twice using a germanium semiconductor detector, and on neither occasion was any trace of cesium found (the limit of detection for cesium-137 is 1.5–2.3 Bq/kg) (Figs. 9.6 and 9.7).

From these findings we can infer that cultivation of rice grains for use as fodder and whole rice plants for silage is likely to be feasible in radiation-contaminated areas.

Chapter 10
Developing and Trialing a System to Monitor Radionuclides in Individual Plots of Farmland to Help Reconstruction Farming in Contaminated Areas

Toshiyuki Monma, Puangkaew Lurhathaiopath, Youichi Kawano, Dambii Byambasuren, Yuta Ono, and Quar Evine

Abstract We believe that if Fukushima Prefecture's agriculture is to be saved, it is essential to create and effectively utilize a system for monitoring the radioactive contamination in each individual parcel of farmland. We are therefore currently developing such a system, as well as a mechanism for putting it to practical use, in the heavily contaminated Tamano district of Soma, adjacent to the village of Iitate. In this chapter, we describe the research outcomes verified to date relating to the characteristics and applications of the monitoring system we are developing.

Keywords Radioactive contamination • Radioactive substance monitoring system • Decontamination

T. Monma (✉) • D. Byambasuren
Department of International Biobusiness Studies, Tokyo University of Agriculture,
1-1-1 Sakuragaoka, Setagaya-ku, Tokyo 156-8502, Japan
e-mail: monma@nodai.ac.jp

P. Lurhathaiopath
Faculty of Life and Environmental Sciences, Tsukuba University,
1-1-1 Tennodai, Tsukuba City, Ibaraki Prefecture 305-8577, Japan

Y. Kawano
Department of Agro-Environmental Science, Obihiro University of Agriculture and Veterinary Medicine, 11 Nishi2sen Inadamachi, Obihiro-shi, Hokkaido 080-8555, Japan

Y. Ono • Q. Evine
Graduate School of Agriculture, Tokyo University of Agriculture,
1-1-1 Sakuragaoka, Setagaya-ku, Tokyo 156-8502, Japan

© The Author(s) 2015
T. Monma et al. (eds.), *Agricultural and Forestry Reconstruction After the Great East Japan Earthquake*, DOI 10.1007/978-4-431-55558-2_10

10.1 Extent of Radioactive Contamination

Some of the earliest research conducted in areas contaminated by the Fukushima nuclear disaster entailed monitoring the distribution of ambient radiation doses and the extent to which the soil had been contaminated. Initially, there were a large number of researchers gathering a variety of measurements, but gradually such measurements came to be released by the administrative authorities, and the independent research was discontinued. The public bodies representing the administrative authorities reported radioactive contamination in each area as shown in Figs. 10.1 and 10.2. Thanks to the release of such data, residents in the disaster zones and the Japanese public in general were able to understand the actual risk presented to them by radiation and take appropriate defensive measures themselves.

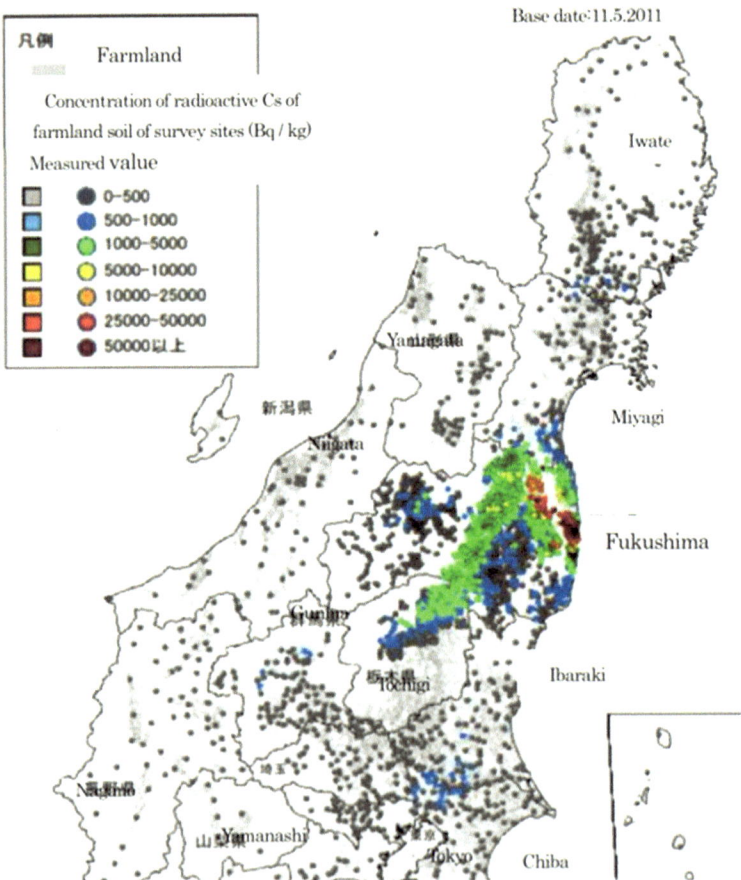

Fig. 10.1 Concentrations of radioactive cesium in farmland soil in the Kanto and Tohoku region

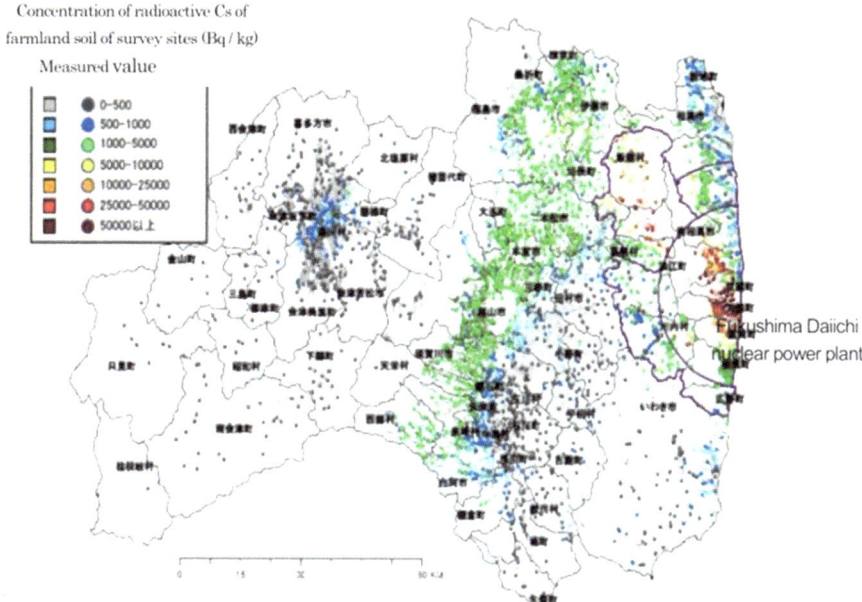

Concentration of radioactive Cs of
farmland soil of survey sites (Bq / kg)

Measured value

	0–500
	500–1000
	1000–5000
	5000–10000
	10000–25000
	25000–50000
	50000 以上

Fukushima Daiichi
nuclear power plant

Fig. 10.2 Concentrations of radioactive cesium in farmland soil in Fukushima Prefecture

The results of these surveys indicated that areas around the Fukushima Daiichi Nuclear Power Station and to the northwest of it were severely contaminated by radiation. At the same time, damage was also found in many other areas including the Aizu Basin; the southern part of Miyagi Prefecture near the border with Fukushima Prefecture; along the border between Iwate and Miyagi Prefectures; northern Tochigi Prefecture; areas around the borders of Ibaraki, Chiba, and Saitama Prefectures; and northern Gunma Prefecture.

10.2 The Importance of a Radioactive Substance Monitoring System in Handling Radioactive Contamination and Reconstructing Agriculture

There are three important issues in response to radioactive contamination: the first is compensation for damage incurred, the second is decontamination of the living environment, and the third is decontamination of the agricultural and forestry production bases. This chapter focuses on the last two issues, relating to decontamination.

Decontamination of the living environment is currently under way to lower ambient radiation doses, so that accumulated doses do not exceed an annual level of 1 mSv and people can live safely in the areas affected. However, decontamination that focuses mainly on the areas where people live and work cannot lower the overall ambient radiation dose in regions where radionuclides are scattered over a wide area.

The agricultural and forestry industry production bases that require decontamination include paddy fields, non-paddy arable fields, pastures, and forests. It is important to note that agriculture and forestry are the main industries in the areas contaminated by radiation. If the recovery of these industries is delayed by the contamination incurred, not only will this cause the industries themselves to decline locally, but it will also exert a major impact on people's lives and the conservation of the natural environment in the areas affected.

Although rehabilitation of agriculture and forestry in the contaminated areas is an urgent issue, the national government has been slow to respond, faced as it is with the need to allocate huge budgets for decontamination and compensation. Meanwhile, the agricultural and forestry industries in these areas are on the verge of a crisis because of a whole array of factors including the contamination itself, the damage caused by negative reputation, the evacuation of farmers to other locations, and the outflow of young people from the remaining habitable areas.

At Tokyo University of Agriculture, we believe that if Fukushima Prefecture's agriculture is to be saved, it is essential to create and effectively utilize a system for monitoring the radioactive contamination in each individual parcel of farmland. We are therefore currently developing such a system, as well as a mechanism for putting it to practical use, in the heavily contaminated Tamano district of Soma, adjacent to the village of Iitate.

In this chapter, we describe the research outcomes verified to date relating to the characteristics and applications of the monitoring system we are developing.

10.3   The Purpose of Developing a Radioactive Substance Monitoring System

Three factors explain why radioactive contamination is severely inhibiting recovery in agriculture and forestry.

1. *Delay in understanding the extent of the radioactive contamination in farmlands, forests, and elsewhere:* Although a more detailed map of ambient radiation doses has now been created based on grid units, the extent to which radioactive contamination has spread within soil, forests, and timber is still not fully understood (Fig. 10.3).
2. *Unease regarding the methods used to remove radionuclides and the effectiveness of the removal.* The challenges of decontaminating extensive farmlands where radionuclide have accumulated are compounded by residents' increasing doubts about the efficacy of the decontamination.

Fig. 10.3 Abandoned agricultural land in Namie

3. *Unease about continued reputation-based damage.* More than 3 years have passed since the explosions at Fukushima Daiichi Nuclear Power Station, and since then the Japanese public's concerns about the safety of Fukushima Prefecture's agricultural products have diminished, but the damage from negative reputation persists unabated. The key to recovering agriculture in Fukushima is to earn consumers' trust with regard to the safety of its agricultural products.

Having understood the full implications of these problems, we have been assisting in the recovery effort, guided by the following approaches.

1. Continue agricultural production in habitable areas without prohibiting crop planting. Otherwise, farmlands will rapidly go to ruin (Figs. 10.3 and 10.4).
2. Our consumer survey results, described later in this book, indicated that the way to ensure peace of mind among the public is not to simply lower the limit for radioactive content in Fukushima's agricultural produce. It is instead essential to ensure that none of Fukushima's agricultural produce distributed in the market ever contains any radionuclide detectable using standard detectors.

Our fundamental approaches to recovering agriculture and forestry in irradiated areas can be summarized thus. To put these principles into practice to bring about tangible recovery, we need to establish a system for each district to monitor radionuclides in individual parcels of cultivated land, matching decontamination measures to specific circumstances. We also need to select and produce safe crops and support the development of new agricultural businesses.

Fig. 10.4 Damage by wild boars in the Tamano district. (From blog of Ohashi industrial president)

10.4 Description of the Location Used to Develop the Monitoring System: Soma's Tamano District

10.4.1 Overview of Tamano

The Tamano district was incorporated into the city of Soma via municipal merger in 1954. Located between central Soma and the prefectural capital, Fukushima City, the Tamano district sits in a semi-mountainous area in the Abukuma Highlands, where the climate is comparatively cold. Tamano includes the four areas of Higashi Tamano, Nishi Tamano, Fukuryozen, and Ryozen. More than 60 % of the residents make their living from farming, one of the district's main industries. Tamano's farming is diverse, including raising dairy and beef cattle as well as cultivation of rice, vegetables, and flowering plants. In recent years, the demographic changes that typify Japan's aging society have become more entrenched, with those over 65 years of age increasing and those under 15 years of age decreasing as a proportion of the farming household population. The problems presented by the farmers' advanced age, their lack of successors, and the resulting increase in abandoned land are now serious (Table 10.1).

10.4.2 Damage to Local Agriculture from Radioactive Contamination

The district of Tamano is located about 50-km from the Fukushima Daiichi Nuclear Power Station, next to the village of Iitate, which was designated an evacuation zone. Tamano's ambient radiation dose is therefore relatively high, although the

Table 10.1 Trends and current circumstances in Tamano's agricultural industry

	1970	1990	2010
Total units (number)	206	153	124
Non-farmers	28	29	48
Farmers	178	124	76
Sales farmers (number)	178	124	47
Full-time farmer	40	25	11
First kind part-time farmers	112	17	8
Second-class part-time farmers	26	82	28
Single management (number)	–	64	32
Rice	–	25	21
Dairy, beef cattle	–	24	7
Poultry	–	6	2
Other	–	9	2
Quasi-single complex management (Number)	–	30	11
Complex management (number)	–	35	4
Percentage over 65 years old (%)	–	21.0	34.1
Percentage 15 years old (%)	27.4	18.9	2.3
Cultivated land area (ha)	366	209	106
Abandoned farmland area (ha)	–	19	176

Source: Census of Agriculture

Table 10.2 Results of survey of ambient radiation doses in Soma

	2011.00		2012.00		2013.00	
Location	Soil	Pavement	Soil	Pavement	Soil	Pavement
Soma city	0.74	0.60	0.53	0.36	0.36	0.24
Nakamura	0.49	0.36	0.32	0.23	0.23	0.16
Ono	0.48	0.38	0.37	0.27	0.25	0.17
Iitoyo	0.39	0.34	0.22	0.18	0.18	0.12
Hachiman	0.72	0.57	0.51	0.34	0.36	0.22
Yamagami	1.03	0.74	0.64	0.41	0.47	0.29
Nitaki	0.55	0.46	0.37	0.27	0.29	0.20
Isobe	0.38	0.28	0.27	0.19	0.20	0.14
Tamano	1.88	1.70	1.56	1.00	0.93	0.60

Source: "Information about radiation." Soma HP

Note: The number in the table is the average value; the unit is μSv/h

annual cumulative dose remains below 20 mSv, so no evacuation order has been issued. Nonetheless, comparatively high radiation levels were detected in some areas and evacuation of the residents was seriously considered. The results of a grid survey of ambient radiation doses across Soma conducted by the city authorities showed that the average ambient dose in the Tamano district was highest immediately after the disaster at 1.88 μSv/h. The dose exhibited a downward trend subsequently, but 2.5 years after the disaster it still remains high at 0.93 μSv/h (Table 10.2).

The damage sustained by local agriculture from radioactive contamination was severe. Some of the rice produced in 2011 was found to exceed the newly set cesium limit of 100 Bq/kg. Although rice production is a key revenue source for farmers in the district, therefore they decided to voluntarily refrain from planting rice in fiscal 2012, and rice planting was postponed until fiscal 2013, when thorough decontamination would be complete.

Furthermore, radiation levels are also high in dairy farming pastures, and farmers are therefore prohibited from feeding the grass to their dairy cattle. To be able to continue farming they now rely on hay bought from other areas.

10.5   Overview of the Radioactive Substance Monitoring System under Development

As already described, it is clear that radioactive contamination is significantly inhibiting the recovery of the agriculture and forestry industries of Tamano. Thus, we are attempting to develop a practical monitoring system to ensure that safe agricultural commodities are produced and shipped.

In specific terms, this aim involved collecting and analyzing basic data to develop a monitoring system that could help us to decide decontamination measures, implement them, and evaluate their effects, for each parcel of farmland. Such basic data included the ambient radiation dose (1 m above ground), the soil surface dose (1 cm from the ground), the concentration of radionuclide in the soil (at 0–5 cm and at 5–10 cm depths), the depth of the topsoil, and the soil characteristics (cation-exchange capacity, exchangeable calcium, exchangeable potassium, available phosphoric acid, total nitrogen, soil acidity, etc.). We started the survey in June 2012, and by September we had collected basic data on 646 parcels (142 ha) of farmland including paddy fields, non-paddy arable fields, pastures, and greenhouses across Tamano. Table 10.3 and Figs.10.5 and 10.6 present the aim of surveying each item, the method used, and state of research; Table 10.4 presents the number of farmland parcels surveyed and their areas. In terms of area, paddy fields, non-paddy arable fields, and pastures accounted for 34 ha, 46 ha, and 62 ha, respectively, amounting to a total of 142 ha. In terms of numbers of parcels, paddy fields, non-paddy arable fields, and pastures accounted for 263 parcels, 278 parcels, and 105 parcels, respectively, making a total of 646 parcels. In the Higashi Tamano and Nishi Tamano areas, the parcels comprised mainly paddy fields and non-paddy arable fields, whereas in the Fukuryozen and Ryozen areas, the parcels comprised mainly pastures and non-paddy arable fields (Table 10.4). Table 10.5 shows part of the database created.

Table 10.3 Items surveyed, aims, and methods

Survey item	Research objectives	Investigation and measurement methods
① Basic information (owner name, growers' name and farmland area, etc.	owner, growers, land use situation, and area in the investigated land	Interviews with local leaders
② Space dose of 1 m (μSv/h)	Grasp the external exposure amount of farmer	Using scintillation survey meter TCS-172B, measured at 1 min at a height of 1 m
③ Radiation dose of the soil surface 1 cm (μSv/h, CPM)	Grasp the radiation dose from the soil	Using a scintillation survey meter TCS-172B·GM survey meter TGS-146B, measured in 1 min at the height 1 cm with lead shielding
④ Depth of the cultivated soil (cm)	Selection of appropriate decontamination method	Measuring the hardness, the depth of cultivated soil by use of the soil penetration meter hand auger
⑤ Radioactive material concentration of each depth of soil (Bq/kg)	Grasp the difference radioactive material concentration each soil depth	The collected samples the soil 0–5 cm, 5–10 cm, measured at 3 min using an auto gamma system AccuFLEXγ7010
⑥ Soil nutritional status	Design the fertilization after decontamination (required nutrition, corrosion content, cation-exchange capacity, etc.)	Measured collecting samples of soil depth of 15 cm

10.6 Results of Radioactive Substance Monitoring and Its Practical Uses

10.6.1 Monitoring Results

10.6.1.1 Concentrations of Radionuclide in Farmlands and Efficacy of Decontamination

Characteristics of Radioactive Substance Concentration in Farmlands

Table 10.6 shows the results of surveying ambient radiation doses and radioactive contamination levels in the topsoil of Tamano's farmlands, categorized by location and land type. For Tamano as a whole, the average ambient dose at a height of 1 m is 1 μSv/h and the surface radiation dose is 0.34 μSv/h. The depth of the farmland topsoil is 17 cm, and the concentration of radionuclide in the soil varies from 2,700 to 5,900 Bq/kg. Among the various locations, the ambient dose is highest in Ryozen,

Fig. 10.5 Devices to measure ambient radiation doses and radionuclides

at 1.15 μSv/h, followed by Nishi Tamano, Fukuryozen, and Higashi Tamano. The surface radiation doses are low in Higashi Tamano and Nishi Tamano, but they are relatively high in Ryozen and Fukuryozen. The depth of farmland topsoil is 19–21 cm in Higashi Tamano and Nishi Tamano and 9–10 cm in Ryozen and Fukuryozen.

Turning to the concentrations of radionuclide within the soil itself, in Nishi Tamano and Ryozen, as well as in the Tamano district as a whole, the levels in the lower layer 5–10 cm below the surface are about half the levels in the upper layer 0–5 cm below the surface. In Higashi Tamano, on the other hand, there is a limited difference in the concentrations between the upper and lower layers, which can be attributed to the effects of incorporating rice straw into the soil and plowing in fiscal 2011. Meanwhile, in Fukuryozen, where the concentration of radionuclide is much higher in the upper layer, the topsoil is shallow, and a large volume of radionuclide was deposited on the upper layer in the pastures, which are not usually plowed.

Fig. 10.6 Measuring ambient radiation doses

Table 10.4 Number of parcels and areas of farmland surveyed

Area		Total	Higashi tamano	Nishi tamano	Fukuryozen	Ryouzen
Number of growers (person)		134	35	61	21	21
Investigated farmland	Field number (plot)	646	167	320	103	56
	Area (ha)	142.42	44.24	46.95	44.11	7.12
Paddy	Field number (plot)	263	92	161	2	8
	Area (ha)	34.3	13.9	19.1	0.1	1.2
Field	Field number (plot)	278	65	136	41	36
	Area (ha)	46.1	27.5	12.6	3.6	2.4
Pasture	Field number (plot)	105	10	23	60	12
	Area (ha)	62.1	2.8	15.3	40.4	3.5

In terms of land type, the pastures register a high ambient radiation dose of 1.19 μSv/h, compared to an ambient dose of 0.94–0.97 μSv/h in paddy fields and non-paddy arable fields. In such paddies and arable fields the concentration of radionuclide in the soil's lower layer is about half that in the upper layer. In the pastures, however, an extremely low level in the lower layer contrasts with an extremely high level in the upper layer. The farmlands surveyed also include many

Table 10.5 Extract from monitoring system database

No.	Area	Place	Field number	Land category	Field size (a)	1 m	1 cm	CPM	Depth of topsoil	Bq/kg (Dry soil: moisture 30%) 0–5 cm	5–10 cm
1	Higashi tamano	Syoubusawa	1–39	Paddy	10	0.892	0.326	256	15	4,229	4,447
2	Higashi tamano	Syoubusawa	1–41	Paddy	10	0.934	0.732	273	10	3,889	3,559
3	Higashi tamano	Syoubusawa	1–42	Paddy	10	0.902	0.346	251	15	2,948	3,309
4	Higashi tamano	Syoubusawa	1–4	Field	3	1.177	0.403	347	0	6,300	3,442
5	Higashi tamano	Syoubusawa	1–8	Field	3	1.080	0.300	310	0	3,960	4,911
6	Higashi tamano	Syoubusawa	1–26	Field	10	0.928	0.308	225	25	1,656	2,492
7	Higashi tamano	Syoubusawa	1–31	Field	20	0.962	0.310	279	30	3,825	3,610
8	Higashi tamano	Tachigaro	1–10 ②	Paddy	15	0.884	0.244	272	15	4,711	3,852
9	Higashi tamano	Tachigaro	1–10 ③	Paddy	5	0.822	0.276	255	20	2,917	3,558
10	Higashi tamano	Tachigaro	1–10 ④	Paddy	5	0.820	0.268	227	15	2,476	2,678
11	Higashi tamano	Tachigaro	1–10 ⑤	Paddy	5	0.826	0.280	238	15	4,282	3,780
12	Higashi tamano	Tachigaro	1–5 ①	Field	12	0.752	0.208	215	30	2,890	2,597

No.	Area	Place	Field number	Land category	Field size (a)	1 m	1 cm	CPM	Depth of topsoil	Bq/kg (Dry soil: moisture 30%)	
										0–5 cm	5–10 cm
13	Higashi tamano	Tachigaro	1–5 ②	Field	–	0.734	0.254	228	30	3,500	2,897
14	Higashi tamano	Tachigaro	1–10 ①	Pasture	15	1.446	0.552	538	10	24,681	44
15	Higashi tamano	Ubagaiwa	10	Field	100	0.818	0.216	216	0		
16	Higashi tamano	Ubagaiwa	23–1, 23–2	Field	120	0.754	0.242	229	0		
17	Higashi tamano	Ubagaiwa	26	Field	3	1.236	0.384	261	15	3,054	4,862
18	Higashi tamano	Ubagaiwa	28–1	Field	170	0.400	0.124	164	30	1,659	3,883
19	Higashi tamano	Ubagaiwa	29	Field	100	1.222	0.460	294	5	12,773	1,968
20	Higashi tamano	Ubagaiwa	30 ①	Field	170	0.440	0.134	155	15	481	656
21	Higashi tamano	Ubagaiwa	30 ②	Field	–	0.632	0.240	187	20	3,319	3,356
22	Higashi tamano	Ubagaiwa	30 ③	Field	–	0.558	0.138	154	25	826	1,031
23	Higashi tamano	Ubagaiwa	30 ④	Field	–	0.814	0.286	192	15	3,899	5,567
24	Higashi tamano	Ubagaiwa	30 ⑤	Field	–	0.958	0.258	272	20	3,220	3,121

(continued)

Table 10.5 (continued)

No.	Area	Place	Field number	Land category	Field size (a)	1 m	1 cm	CPM	Depth of topsoil	Bq/kg (Dry soil: moisture 30%) 0–5 cm	5–10 cm
25	Higashi tamano	Ubagaiwa	30 ⓖ	Field	–	0.566	0.356	148	20	3,332	1,989
26	Higashi tamano	Ubagaiwa	31	Field	170	0.341	0.103	107	0		
27	Higashi tamano	Ubagaiwa	32	Field	170	0.386	0.114	119	25	1,675	174
28	Higashi tamano	Ubagaiwa	33	Forest	170	0.368	0.098	99	30	1,350	836
29	Higashi tamano	Ubagaiwa	35	Field	170	1.022	0.246	246	0		
30	Higashi tamano	Ubagaiwa	36	Field	50	0.966	0.276	246	0		
31	Higashi tamano	Shigekari	1–177	Field	100	0.714	0.196	180	15	2,251	2,398

Table 10.6 Ambient radiation doses of farmlands and concentrations of radionuclide in the soil

		Air dose (ground 1 m)	Soil surface dose (1 cm)	Depth of topsoil	Concentration of radioactive substances in soil (Bq/kg)	
		(µSV/h)	(µSV/h)	(cm)	0–5 cm	5–10 cm
By area	Average of area	1.00	0.34	17	5,933	2,708
	Higashi Tamano	0.86	0.26	19	4,045	3,233
	Nishi Tamano	1.06	0.28	21	5,876	2,842
	Fukuryozen	0.98	0.68	10	5,706	1,270
	Ryozen	1.15	0.33	9	7,968	3,063
Type of land use	Paddy	0.94	0.25	21	5,008	2,757
	Field	0.97	0.27	14	5,113	3,330
	Meadow-pastureland	1.19	0.70	7	8,715	1,001

Note: Concentration of radionuclide is a converted value of 30 % moisture

hotspots with a high concentration of radionuclide in their upper layer, exceeding 10,000 Bq/kg, particularly in pastures, as well as in unplowed paddy fields and non-paddy arable fields. It is essential to establish a risk management system using radioactive contamination maps in these areas.

Decontamination of Paddy Fields and Its Efficacy

These survey results were used to consider possible decontamination measures for the farmlands in Tamano. The specific measures chosen for paddy fields and non-paddy arable fields involved deep plowing and scattering soil improvement agents. In pastures, on the other hand, the shallow topsoil meant that just a thin layer of surface soil would be removed, with soil brought in from other areas where necessary. Decontamination work started in Tamano's paddy fields at the end of November 2012. However, the arable soil in Tamano's paddy fields is generally not very deep, and many areas contain rocks underneath the soil, so the work was undertaken carefully at 1.5 times the normal depth, one paddy at a time. Large tractors were not used to avoid damaging the plow sole. Cesium absorption was inhibited by scattering 200 kg zeolite and 50 kg potassium chloride per 10 ares.

In non-paddy arable fields, individual farmers undertook decontamination and started planting crops. Decontamination of pastures started in July 2013, and sowing of grass is planned for that fall.

We first evaluated the efficacy of Tamano's paddy field decontamination in May 2013. The survey results are compiled in Table 10.7. Looking first at the ambient radiation dose before decontamination (in July 2012) and after decontamination, we find that it dropped by about 0.1 µSv/h from 0.83 to 0.73 µSv/h in Higashi Tamano and by about 0.15 µSv/h from 1.01 to 0.86 µSv/h in Nishi Tamano. However, it proved impossible to meet the Japanese Ministry of the Environment's target of

Table 10.7 Ambient radiation doses and concentrations of radionuclide in paddy field soil in Tamano before and after decontamination

Before decontamination (2012)	Air dose (μSv/h)	Concentration of soil radioactive substances (Bq/kg)			
		0–5 cm		5–10 cm	
Higashitamano	0.83	4,050		3,312	
Nishitamano	1.01	5,576		2,423	
After decontamination 2013	Air dose (μSv/h)	Concentration of soil radioactive substances (Bq/kg)			
		Cs total	Cs-137	Cs-134	K-40
Higashitamano	0.73	3,213	2,069	1,144	799
Nishitamano	0.86	4,107	2,651	1,456	1,010

Note: Concentration of radionuclide is a converted value of 30 % moisture

lowering the ambient radiation dose by half, indicating that it is difficult to lower ambient radiation doses in semi-mountainous areas surrounded by mountain forests.

Moving on to changes in the concentrations of radionuclide in the soil, in 2012 we measured the soil contamination levels in the paddy fields at two different depths, 0–5 cm and 5–10 cm. In fiscal 2013, however, the paddies had been deep plowed during the decontamination process, so we took soil samples for measurement without making distinctions based on depth. As a result of the soil having been mixed, the total cesium concentration was found to have decreased by about 1,000 Bq/kg compared to the concentration in the 0–5 cm layer in 2012.

Figure 10.7 summarizes the efficacy of decontamination in each parcel of paddy field in the form of a map. As the map clearly shows, cesium concentration in many of the paddies decreased. However, upon closer examination, there are also some paddies in which the concentration did not decrease, indicating that it is essential to investigate why this occurred and to study how to inhibit absorption of radionuclide.

Decontamination of Meadows and Grazing Land and Its Efficacy

Radionuclide in pastures are accumulated primarily on the surface. To decide how to decontaminate the pastures, we examined data we collected for monitoring system development purposes. The data included the depths of the surface soil, the concentrations of radionuclide, and the ambient radiation doses for each parcel of meadowland and grazing land (Table 10.8).

The specific decontamination method we selected was use of a backhoe to scrape off 4 cm of surface soil in meadows and grazing land. This decontamination work is currently under way, with completion scheduled for the end of October 2013. Although the efficacy of the decontamination will be evaluated in detail later, the concentrations of radionuclide measured in some of the meadows and grazing land

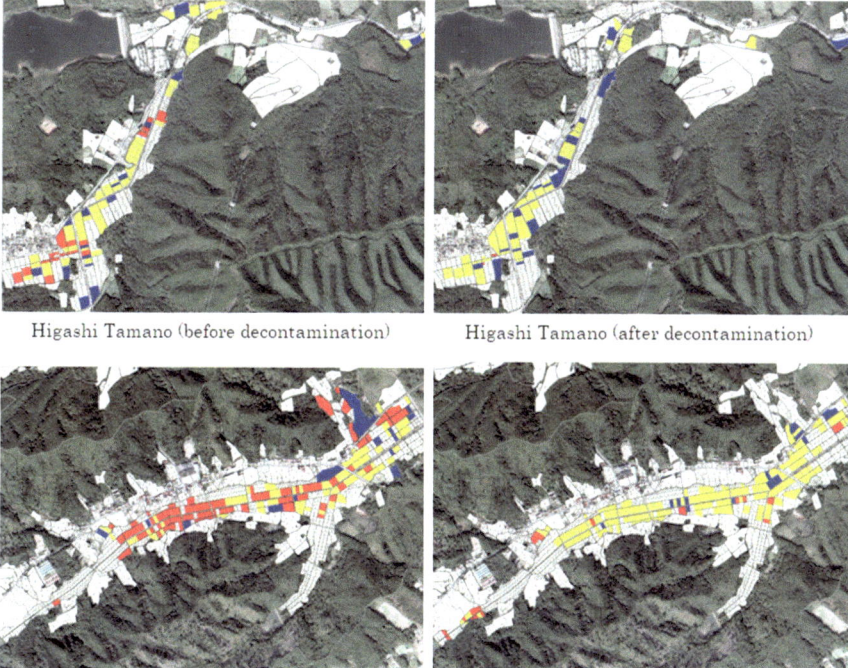

Higashi Tamano (before decontamination)   Higashi Tamano (after decontamination)

Nishi Tamano (before decontamination)   Nishi Tamano (after decontamination)

Fig. 10.7 Partial map of radioactive contamination in paddy field soil in Higashi Tamano and Nishi Tamano before and after decontamination. Note: Colors in the figure are classified by concentration of radioactive soil: *blue*, less than 3,000 Bq/kg; *yellow*, 3,000–4,999 Bq/kg; *red*, more than 5,000 Bq/kg

Table 10.8 Ambient radiation doses and concentrations of radionuclide in meadows and grazing land (2012)

Location	Air dose (μSv/h)	Concentration of soil radioactive substances (Bq/kg)	
		0–5 cm	5–10 cm
Nishitamano	1.26	6,769	1,021
Fukuryozen	1.09	8,604	895
Ryozen	1.31	9,480	2,279

Note: Concentration of radionuclide is a converted value of 30 % moisture

that have already been decontaminated were found to have dropped to one-tenth of their original levels.

However, the migration of radionuclide from the soil to the grass remains an issue. We investigated the migration of radionuclide to the first grass of 2013 for each parcel of meadow and grazing land. As shown in Table 10.9, the migration coefficient differed substantially for each location. In research conducted previously,

Table 10.9 Concentrations of radionuclide in meadows and grazing lands and coefficients of migration to grass

Location	Air dose (µSv/h)	Concentration of soil radioactive substances (Bq/kg)			
		Cs total	Cs-137	Cs-134	K40
Nishitamano	1.57	5,858	3,776	2,081	1,339
Fukuryozen	0.94	5,953	3,836	2,117	1,438
Ryozen	1.30	11,922	7,732	4,190	2,713
	Concentration of soil radioactive substances (Bq/kg)				Transition coefficient
	Cs total	Cs-137	Cs-134	K40	
Nishitamano	67	38	29	158	0.014
Fukuryozen	239	147	92	217	0.042
Ryozen	708	445	263	325	0.058

Note: Concentration of radionuclide is a converted value of 30 % moisture

the generally accepted coefficient for migration of radionuclide to grass had been 0.045. The discrepancies that appeared in our research could be explained by differences in soil types, grasses planted, and manure used in the meadows and grazing lands.

10.7 Future Issues and Trends in the Use of Radioactive Substance Monitoring Systems

The monitoring system we developed has been actively used in decontaminating farmlands. Decontamination of paddy fields has already been completed, and the decontamination of meadows and grazing lands is currently in progress. In addition to its use in publicly funded decontamination of farmlands, the monitoring system has also been actively used by individual farmers, who familiarized themselves with the concentrations of radionuclide in their own farmlands to adopt their own countermeasures. We therefore provided the data for each parcel of farmland in the form of feedback to all the owners and cultivators. In addition, the map of radioactive substance concentrations in farmlands across the entire area was provided to the chiefs of the Higashi Tamano, Nishi Tamano, Ryozen, and Fukuryozen areas, and was also made available to all residents by displaying it at assembly halls and other public buildings in each area. Ideally, the monitoring system should be used by the farmers themselves to monitor radionuclide in each area as a whole. We therefore need to structure the system so that the farmers can measure such substances simply, rapidly, and accurately. Therefore, we estimated the cost of the monitoring system development (Lurhathaiopath et al. 2014) and proposed the direction of agricultural policy in the radioactive contamination area.

References

Lurhathaiopath P, Kawano Y, Monma T (2014) Farming reconstruction support and the development of each farmland radioactive materials monitoring system. J Farm Manag Soc Jpn 52(1,2):67–72 (in Japanese)
Monma T (2014) Radioactive contamination and agricultural policy. J Rural Soc Econ 32(1):15–24 (in Japanese)

Chapter 11
New Decontamination Methods for Parks and Other Areas in Which Radionuclides Have Accumulated

Mitsuo Kondo and Chizuko Mizuniwa

Abstract Activities to remove radionuclides in the wake of the Fukushima Daiichi nuclear accident were unprecedented in nature and required the utmost urgency. The methods employed generally involved pulling plants out by the roots. Arguably, however, such drastic measures were unavoidable to achieve the paramount objective of reducing radiation dosage, thus avoiding irradiation of residents and ensuring safety. However, pulling plants out by the roots can have extremely harmful consequences. Not least of these is the volume of waste (trunks, branches, leaves, roots, soil, etc.) generated and the way in which the waste is disposed. In addition, the land is stripped bare by the decontamination process, compromising the various functional benefits the plants had previously provided to the community. Our approach to decontaminating trees and wooded areas is aimed at improving the effectiveness and efficiency of the decontamination process by utilizing current landscape architecture knowledge and techniques.

Keywords Decontamination • Parks • Grass fields • Wooded area • Radiation blocking

11.1 Introduction

Activities to remove radionuclides in the wake of the Fukushima Daiichi nuclear accident were unprecedented in nature and required the utmost urgency. The methods employed generally involved pulling plants out by the roots. Arguably, however, such drastic measures were unavoidable to achieve the paramount objective of reducing radiation dosage, thus avoiding irradiation of residents and ensuring safety.

However, pulling plants out by the roots can have extremely harmful consequences. Not least of these is the volume of waste (trunks, branches, leaves, roots,

M. Kondo • C. Mizuniwa (✉)
Department of Landscape Architecture Science, Tokyo University of Agriculture,
1-1-1 Sakuragaoka, Setagaya-ku, Tokyo 156-8502, Japan
e-mail: mizuniwa@nodai.ac.jp

© The Author(s) 2015
T. Monma et al. (eds.), *Agricultural and Forestry Reconstruction After the Great East Japan Earthquake*, DOI 10.1007/978-4-431-55558-2_11

soil, etc.) generated and the way in which it is disposed. In addition, the land is stripped bare by the decontamination process, compromising the various functional benefits the plants had previously provided to the community.

11.2   Decontamination of Greenery: Basic Policy and Specific Initiatives

The authors' fundamental approach toward the decontamination of green areas is to fully utilize landscape architecture knowledge and techniques to achieve the dual objectives of decontaminating the area while preserving greenery and generating as little waste as possible. To achieve this, we must first ascertain the distribution of radionuclide and determine the parts of the greenery that need to be removed. Then we can implement effective decontamination measures as required, including cropping or pruning away the contaminated sections. Furthermore, we aim to renew and revitalize plants through these processes, improving their appearance, health, and functionality compared to their pre-decontamination state.

In this chapter we describe the outcomes of a series of practical experiments and field tests we conducted to develop techniques to use at contaminated sites. The four decontamination methods we have devised to date are (1) landscape architecture-based lawn decontamination and renewal (a method of decontaminating Japanese lawn grasses by removing the grass protruding above ground while preserving the underground stolons and root system, followed by application of topdressing and fertilizer to renew and revitalize the grass); (2) radiation blocking (a method of reducing radiation dosage by covering contaminated areas such as lawns with a thick top layer of grass or another material, without removing the existing plant life); (3) radiation capture (a method of preventing recontamination and secondary contamination in risk areas by using plants such as mosses to capture and absorb radionuclide); and (4) radiation reduction (a method of absorbing and partially removing radionuclides from the soil using greenery such as grasses that can regrow after mowing).

11.3   Landscape Architecture-Based Lawn Decontamination and Renewal Method

Rather than pulling out warm-climate grasses such as mascarene grass and noshiba (*Zoysia japonica* Steud.) at the roots, only the parts in which radionuclides have accumulated are removed: this includes removing the stems and leaves, matted cut grass clippings and erect stems accumulated on the soil surface, and stolons protruding above ground. Alternatively, we remove the top 1–2 cm of soil while preserving the underground stolons and root system. Subsequently, topsoil and fertilizer are applied to promote new growth, renewing and revitalizing the grass. For

Fig. 11.1 Practical experiment using a field top maker to decontaminate a park lawn area

extensive areas of lawn such as green spaces in parks, large vehicle-mounted turf strippers (Fig. 11.1) or sod cutters can be used to perform the deep mowing process, whereas domestic string trimmers or lawnmowers can be used for smaller areas such as home lawns.

The major benefit of this method is that the total cost is approximately one quarter that of methods that involve stripping the grass by its roots and replanting following decontamination. Furthermore, the amount of waste generated is reduced by half.

11.4 Radiation Blocking

In this method, rather than stripping off the contaminated grass layer, the surface is covered with either another thick layer of grass (turf cut approximately 10 cm deep) or a thick layer of sprayed vegetative base material. The latter method involves using a specialized compressed air device to spray areas such as bedrock or embankments made of sprayed mortar that lack a substrate to support plant life. Such surfaces are sprayed to a depth of 5 cm or more with a mixture of organic vegetative base material (a combination of bark compost, peat moss, etc.) combined with fertilizer (slow-acting) and bonding agent (polymer resin, or sometimes bentonite or other substances to increase the radiation-blocking effect), into which grass seed such as tall fescue is mixed. The main advantage of this method is that it generates no waste and can be used in any environment.

11.5 Radiation Capture

Radiation capture is a method whereby cesium or other radionuclides are proactively captured and absorbed to prevent recontamination or secondary contamination caused by wider dispersal. Mosses grown in sheet form are spread on structures such as building walls or water channels to capture radiation in the air or rainwater. Alternatively, aquatic plants are grown at the water's edge to prevent contamination of lakes and marshes from inflow of earth and sand containing the radionuclide. Our attention focused on mosses in particular as related studies had shown that they have a strong tendency to accumulate radionuclides. We decided to turn this capacity to good use, and came up with this method of using mosses to capture radiation.

The new method requires cultivated sheets of moss to be fixed onto flat or slanted rooftops, or the walls of buildings. The sheets are also attached to areas where radionuclides tend to accumulate, such as drain holes and gutters, and can be wrapped around the trunks of trees at the roadside or in gardens and parks. The method utilizes the capacity of moss to proactively capture radionuclide in the air and rainwater. After a certain time, the sheets attached to or wrapped around an object are removed, and new sheets applied; this helps prevent secondary contamination and recontamination, and also lowers the ambient radiation dose to a certain extent.

Next, we studied eight major types of aquatic plants grown in the exhibition garden at the city of Koriyama's Ouse Park, measuring and comparing their ability to adsorb and accumulate radionuclide. This approach enabled us to evaluate their potential for use as radiation capture agents in waterside areas. As a result, we found that horsetail and Japanese rush plants contained a high concentration of radioactive material. In addition, their relative ease of propagation and rapid growth make them highly effective candidates for use as radiation capture agents in waterside areas.

11.6 Radiation Reduction (Absorption and Removal of Radionuclides in Soil Through Phytoremediation by Greening Plants)

We set out to determine what characteristics make plants potential candidates for use in absorbing radionuclide. According to our own current list of requirements, a candidate plant should (a) be evergreen, (b) be capable of propagation from seed, (c) have a high germination rate, (d) have easily obtainable seeds, (e) have stalks and leaves that cover a large ground area, (f) be able to withstand poor soil conditions, (g) be hardy and able to withstand neglect, (h) be able to help maintain or improve soil fertility, (i) be able to withstand harsh heat, (j) be fire resistant, (k) not have flowers that attract birds or insects (to prevent the spread of radiation via these agents), (l) have absorptive roots concentrated near the soil surface, (m) be able to absorb and accumulate radioactivity year after year, (n) have foliage that does not protrude too far from the ground and does not generate much bulk when trimmed and collected, and (o) be an energy-generating plant that can be used as biomass.

From the authors' observations so far, ideal candidate grasses include dwarf tall fescue and Bermuda grass, which have conventionally been used to green sloping areas and to create evergreen lawns and can quickly regrow after being cut. We are currently conducting cultivation tests to determine their ability to absorb cesium.

11.7 Decontamination of Trees

11.7.1 Mechanism by Which Trees and Wooded Areas Are Contaminated

Radionuclides (isotopes) scattered aerially in the event of a nuclear accident contaminate trees and woodland areas through a variety of routes and mechanisms. This contamination occurs partly through direct routes whereby radiation enters the tree after attaching to the leaves and branches of the tree crown or to the trunk, and it also occurs partly through indirect routes whereby radiation is deposited on or within the soil and is subsequently absorbed by the plant. Trees are either evergreen or deciduous, and the mode of contamination varies accordingly.

In the case of the Fukushima nuclear accident, the trees were irradiated around the middle of March, so deciduous trees were still at the winter bud stage and had not yet developed new shoots. Thus, the new leaves that subsequently developed were not exposed directly to the radioactive fallout. Nonetheless, it has been reported that the fallen leaves showed a high concentration of radionuclide that year. It is therefore hypothesized that radionuclide that attached to the winter buds may have transferred directly to the new leaves that subsequently developed, and that some of the radiation absorbed by the roots from the soil may also have transferred to the leaves.

Meanwhile, in evergreen trees it is possible for radionuclides to be detected in new leaves that grow on the irradiated branches later in the same year. Evidence of this was provided when radionuclides such as cesium were detected in tea leaves grown in some parts of Shizuoka's tea-growing areas when the first tea was picked in 2011, leading to a temporary ban on shipments. These observations imply that it may be possible for radionuclide attached to the old leaves to be absorbed into the plant then transferred internally to the new leaves. The same phenomenon has, moreover, been observed in fruit, with shipment of persimmons being banned because of detection of radiation. This finding suggests that the same mechanism may apply in other tree varieties.

11.7.2 Overall Approach to Decontamination of Trees and Wooded Areas

Our approach to decontaminating trees and wooded areas is aimed at improving the effectiveness and efficiency of the decontamination process by utilizing current landscape architecture knowledge and techniques. This enables us to develop methods based on a sound understanding of the mechanisms and modes of contamination involved, obviating the need to uproot trees or plants. During this process we aim to aggressively prune overgrown branches and leaves that have not been properly tended to improve the health and vitality of the individual trees and wooded areas as a whole without negatively impacting their appearance.

11.7.3 Specific Methods for Decontamination of Trees

To reduce the tree's total radiation dosage, the overall volume of branches and leaves should be reduced significantly by cutting back or thinning out, without harming the tree's shape or appearance. Any radiation that attaches to the surface soil surrounding the tree and its roots must be thoroughly removed.

The process involves (1) pruning to remove all leaves and branches in the treetop portion of the tree crown (i.e., that year's branches), (2) pruning to thin out leaves and branches in the tree crown to around one third of their original volume, (3) thoroughly washing the branches and trunk of the tree crown with a high-pressure water blaster, or peeling/grinding the bark, (4) removing all fallen leaves under the tree crown, and (5) removing the top 2–3 cm of soil under the tree crown. In steps (4) and (5), the aim is also to collect the radionuclide contained in the water used for blasting and the peeled/ground bark. Following initial decontamination, the situation should be reassessed every 1 to 2 years, with any of steps (1) to (5) repeated as necessary as maintenance work in response to subsequent changes in the tree's radiation dosage and growth.

11.7.4 Decontamination of Bark by High-Pressure Water Blasting or Peeling/Grinding

As are the trees' leaves, the trunks and branches are also contaminated with radionuclide and, naturally, they also require decontamination. However, decontamination methods differ according to the condition of the bark covering the tree surface. Moreover, the condition of the bark varies dramatically depending on the species of tree and also changes as the tree ages.

Although smooth bark is comparatively simple to decontaminate with a high-pressure water blaster, many species of trees have horizontal or vertical wrinkles or cracks in their outer layer of rough bark (which comprises cork or dead phloem tissue that protects the living tissue inside). There may be radionuclide inside these wrinkles or cracks, and as water blasting alone is insufficient to completely remove such contamination, in some cases measures such as peeling/grinding of the outer bark are required. However, peeling/grinding the outer bark raises the risk of damaging the internal tissue, and a careful approach is therefore necessary. The

Fukushima Agricultural Technology Centre has already demonstrated that peeling/grinding the outer (rough) bark of fruit varieties such as grape vines, *nashi* (Asian pear), apple, and persimmon trees proved successful in reducing contamination to a significant extent.

11.8   Forest Decontamination Initiatives

When decontaminating forests, as a general rule areas within 20 m of populated (residential) areas should be treated, either by removing fallen leaves from the forest floor, cutting undergrowth, and removing the A_0 (litter) layer, or by radiation blocking using the thick-base spray method. In other areas, the radiation dose should gradually be reduced by thinning suppressed (overcrowded) trees and controlling density to ensure that the forest is preserved and further cultivated, in addition to caring for and maintaining the forest in other ways, such as by cutting undergrowth.

As part of our research, we attempted to quantify the efficacy of decontamination measures in the Japanese cedar groves and broad-leafed deciduous wooded areas of Ouse Park in Koriyama, Fukushima Prefecture. We removed fallen leaves, forest floor undergrowth, and the A_0 (litter) layer, then studied the rate of decline in the radiation dose and analyzed the distribution of the total radiation dose using a gamma camera. The foregoing measures were found to reduce radiation levels by 10–20 % compared to the levels before decontamination. In light of the fact that felling trees to decontaminate forest areas leads to loss of the functions provided by greenery and generates problems of the large volume of waste generated, there is a need for detailed site studies into the extent to which thorough cleaning of the forest floor is effective in decontamination.

11.9   Establishing Monitoring Methods to Conduct Scientific and Economically Efficient Decontamination

Two years on from the nuclear accident, the dynamics of radioactive contamination are gradually being understood. For the decontamination process to be performed effectively, efficiently, and without waste, and to accurately assess the resulting benefits, radioactive contamination should ideally be mapped in a visible format to give an overview of the situation. This method would allow all parties to share a common understanding of the radiation dose preceding decontamination, and also serve as a decision-making tool when deciding which areas to prioritize, making it easy to check how effective the decontamination process has been and whether there is any unevenness in results. To study the feasibility of this method, we are currently evaluating the performance of gamma cameras and radiation monitors. For some time we have also been using a bio-imaging analyzer to conduct tests at the laboratory level to identify which parts of plants are contaminated. An alternative application for this research could be in managing the safety of temporary and "pre-temporary" storage sites for radioactive waste generated from the decontamination process.

For both decontamination case studies and post-decontamination surveys our aim is to use gamma cameras to map radiation doses in a visible format, as well as radiation monitors to make the measurement process more efficient. Accordingly, we tried out new cameras and monitors and evaluated their performance in on-site tests at parkland green spaces in the cities of Koriyama in Fukushima Prefecture and Matsudo in Chiba Prefecture. As a result of these tests, we found that measurement by gamma camera was an extremely effective tool in mapping the radiation dose visibly, particularly when it came to detecting contamination hot spots in wooded and grassy areas. However, these cameras are extremely expensive, and to obtain accurate measurements several experienced staff are required to operate them and analyze the data. There are, moreover, several other hurdles to using gamma cameras at decontamination sites, including the fact that the measurement process itself takes a significant amount of time. The radiation monitors, on the other hand, are lightweight and portable, and can provide real-time measurements in series as the operator walks around. These monitors proved to be highly effective measurement devices for detecting pre-decontamination hotspots and evaluating post-decontamination benefits.

11.10   Measuring the Directional Contributions to the Ambient Radiation Dose to Perform Effective Follow-Up Decontamination

In the grounds of some residential buildings, ambient radiation doses fail to reduce significantly despite decontamination, which is often caused by the effect of radiation from nearby woodland areas or fields. In such cases, the planning of effective follow-up decontamination measures would be facilitated if we knew the extent to which radiation originating from a certain direction was contributing to the total dose at a particular site. To this end, Michinori Mogi, who collaborated with the authors on this project, developed a collimator capable of dividing a space into six portions. The directional contributions to a radiation dose can be measured by rotating the open face.

11.10.1   Overview of the Apparatus

We created a collimator made of lead 3 cm thick that is capable of housing the entire detection unit of an NaI scintillation spectrometer (AT6101 NaI(Tl), $\varnothing 40 \times 40$ mm). The only opening was a solid angle $2/3\pi$ (sr) from the center of the detection unit, and rays of radiation could reach the detection unit from the direction of the opening uninterrupted. By rotating this open face, the contributions to the total radiation dose from six different directions could be measured.

11.10.2 Sample Measurements

We took measurements in the village of Iitate, Fukushima Prefecture, behind a house that backed onto a hill slope. The hill slope began just 2 m from the house, and we selected a point at a height of 1 m above the middle of the flat area to take measurements. We suspected that the hill slope was contributing significantly to the radiation dose (Fig. 11.2).

First, we measured the background dose by using a 3-cm-thick lead sheet to completely cover the opening of the collimator, which was pointed upward. Compared to an ambient dose of 0.65 μSv/h when the opening was unobscured, the background dose was 0.051 μSv/h. Next, we measured the dose from six different directions, starting from the direction facing the house wall. Measurements were taken before and after obscuring the hill face with sprayed soil (radiation blocking) and decontaminating the flat area by removing the top layer of soil. Measurement results are shown in Fig. 11.3. The sum of the dosage rates minus the background dose from all six directions matched the total unobscured ambient dosage rate minus the background dose to within a 10 % margin of error. Looking at the directional contributions before decontamination, the contribution from the ground below the measurement point was largest, comprising 30 % of the total dose. The contributions from the hill face and the north and south directions comprising the flat area each accounted for 20 %. The contributions from the upward direction and the direction of the house were small, at less than 6 %. Following decontamination and other measures, the overall ambient dose declined to 54 % of the original figure, with the largest reductions (to 37–54 %) witnessed from the directions of the flat

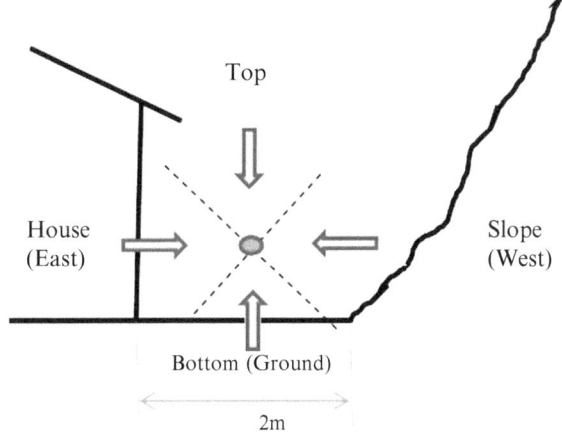

Fig. 11.2 Point at which directional contributions to ambient radiation dose were measured

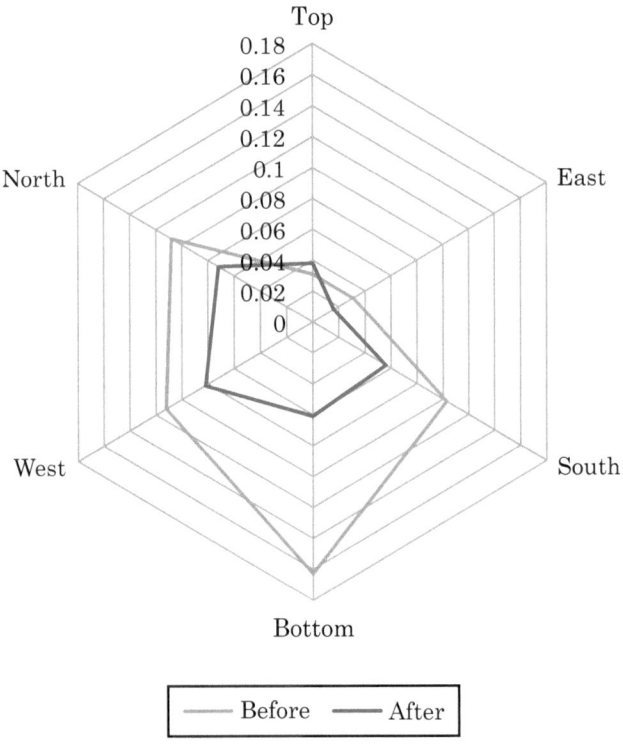

Fig. 11.3 Directional contribution ratio: measurement results (μSv/h)

ground surface where the top layer of soil had been removed. If the house direction and the upward direction are excluded, it is evident that post-decontamination radiation dosages from the other four directions were all lower and their contributions to the dosage at the measurement point became more evenly balanced.

11.10.3   In Conclusion

In the past, we had only an intuitive understanding of directional contributions to radiation doses, but this study enabled us to ascertain contribution ratios quantitatively using a collimator. We believe that application of this method will allow the sources and directions of the strongest radiation doses to be identified, enabling appropriate countermeasures to be taken.

Chapter 12
Forest Problems

Takahisa Hayashi

Abstract Following the Fukushima-Daiichi Nuclear Power Plant accident in March 2011, large amounts of radionuclides were dispersed over the 50-km tract of land between the plant and the city of Date, where forested mountains make up approximately 90 % of the total area. During the weeks after the disaster, the fallout of radionuclides was deposited on trees and local residences in an aerosol or gaseous form partly dissolved in rainwater or snow. Radiocesium was incorporated into plant bodies through all the exposed surfaces, not only through the leaves but also via the bark. Trees, which can directly incorporate radionuclides, serve as one of the largest biological sinks of fallout radionuclides. Thus, they are capable of reducing the harmful impact of these radionucleotides on residents in rural land and others in Japan. The forest in the rural land became in an enormous biological sink of ^{137}Cs with a long half-life, 30.1 years.

Keywords Radiocesium • Migration • Decontamination trials • Social problems

12.1 Introduction

Fukushima Prefecture contains about 970,000 ha of forests with an estimated growing stock of about 100 million m^3. We are surveying and conducting research on forested areas belonging to Soma's regional forestry cooperative in the town of Shinchi, as well as the cities of Soma and Minamisoma. According to SPEEDI (the System for Prediction of Environmental Emergency Dose Information), a large quantity of radionuclides flowed from the Fukushima Daiichi Nuclear Power Station toward the city of Fukushima via the western parts of Minamisoma and Soma. The radionuclides lingered and accumulated when they reached the foothills of the area's mountains, which are more than 400 m in altitude, increasing the amount of radiocesium clinging to the outer bark of trees in forested areas nearby.

T. Hayashi (✉)
Department of Bioscience, Tokyo University of Agriculture,
1-1-1 Sakuragaoka, Setagaya-ku, Tokyo 156-8502, Japan
e-mail: t4hayash@nodai.ac.jp

T. Monma et al. (eds.), *Agricultural and Forestry Reconstruction After the Great East Japan Earthquake*, DOI 10.1007/978-4-431-55558-2_12

179

When we started our research we aimed to investigate the behavior of radioiodine and radiocesium. Radioiodine is known to bind to the woody component in trees, and we therefore investigated the bonding mechanism at the molecular level. We found that the bulk of this radioiodine is taken in through the lenticels of the trees and then captured via bonding with the cell walls. Radiocesium, on the other hand, is absorbed through the leaves and bark, permeating to the wood (the part that will be used as lumber) from the inner bark. Some of the radiocesium also reaches the interior, made up of old tree rings (Fig. 12.1). Since the Chernobyl nuclear disaster in 1986, many papers have been published on the infiltration of radionuclides into forest trees, but the mechanism of the infiltration is still unknown. Although radiocarbon (^{14}C) is taken in and captured in the tree ring of the specific year, radiocesium is found to migrate to the corewood.

The radioiodine and radiocesium generated by the Fukushima Daiichi nuclear plant accident were incorporated into the trees over a vast area of mountain forests, which resulted in protecting humans, we hope. Although radioactive iodine-131 (^{131}I) has disappeared as a result of its short half-life, radiocesium-137 (^{137}Cs) in particular, has a long half-life of 30.17 years, and will therefore remain in the trees: this will turn the forests into enormous unnatural repositories of ^{137}Cs.

The amount of cesium incorporated into the trees in Fukushima Prefecture varies among areas. Here we discuss the relationship between humans and trees as revealed by analysis of the behavior of radiocesium within the trees.

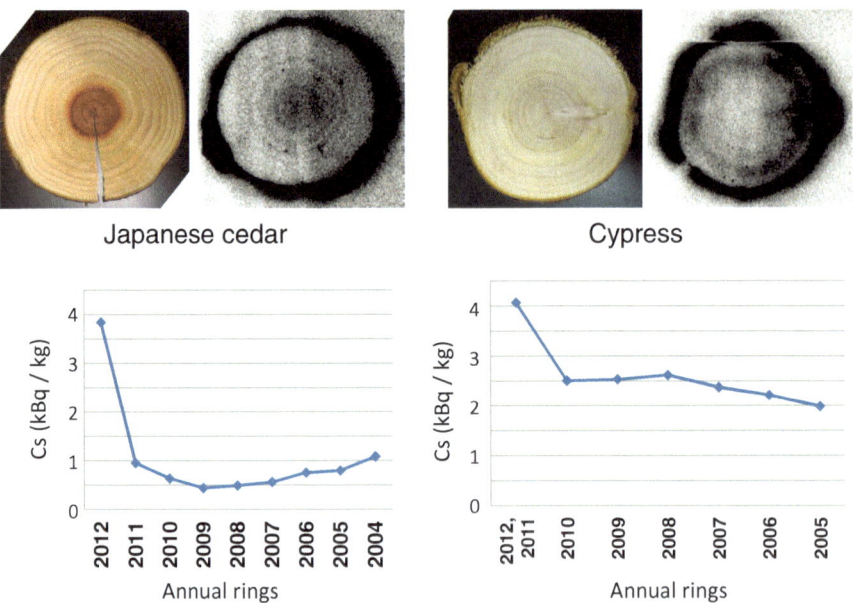

Fig. 12.1 Transverse sections of stems with autoradiographs and levels of radiocesium in the forest trees

12.2   Study on the Migration of Radionuclides Within Forests

We conducted a survey of radionuclide contamination in the forested areas of Minamisoma's Haramachi district. We divided the trees into leaves, branches, bark (outer and inner), and wood (sapwood and heartwood), and measured the radiation in each of these parts.

We also dug trees up and measured how much radionuclide had migrated to the roots (Figs. 12.2 and 12.3), although we could only dig up four trees in a day, even with ten people working from dawn till dusk. As a result of our perseverance, however, we were able to confirm that radiocesium incorporated into the trees from aboveground was infiltrated within the trees and migrated to the roots, from which some was secreted into the soil.

Figure 12.4 summarizes the distribution of radiocesium in forested areas and trees. Much of the cesium accumulated in the litter layer, and only 1 cm of the soil surface was contaminated. We found that most of the radionuclides were captured by the leaves and bark of the trees, as we had expected. However, the radiocesium had also migrated to the wood and roots of the trees to a considerable degree.

Thus, the forests of Fukushima have become enormous unnatural repositories of ^{137}Cs, in which the cesium is circulated to and from the trees via various living organisms (mushrooms, moss, ferns, insects, animals, etc.). We call this the cesium cycle (Fig. 12.5).

Fig. 12.2 Digging a tree up to measure radionuclides in the roots

Fig. 12.3 The dug roots of a tree

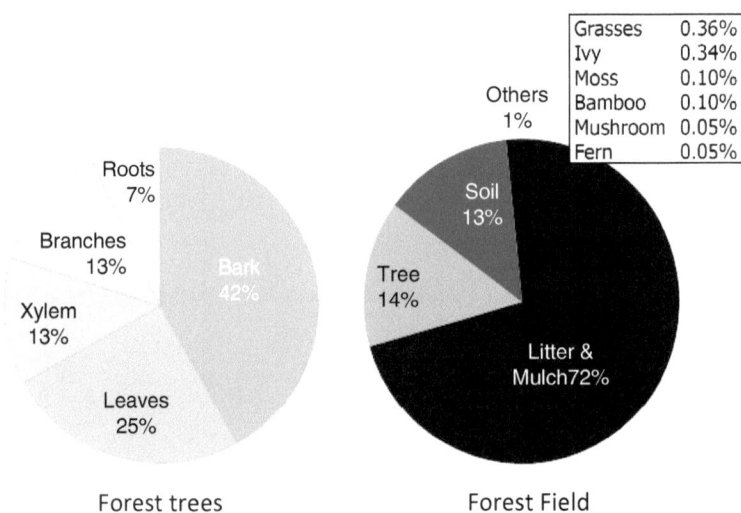

Forest trees   Forest Field

Fig. 12.4 Distribution of radiocesium in the forest in April after the nuclear plant accident

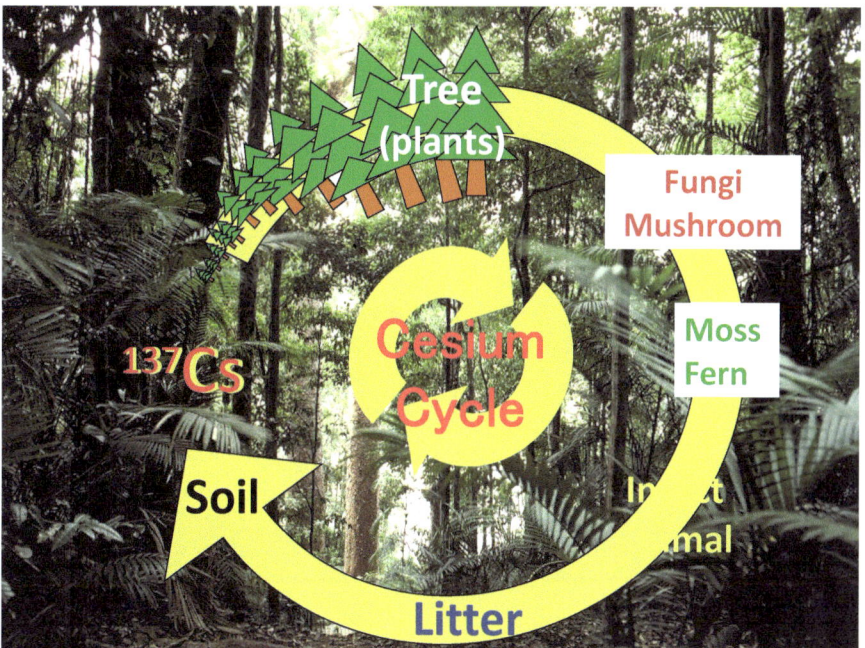

Fig. 12.5 Cesium cycle in the forests of Fukushima

12.3   Decontamination Trials

We conducted an experiment using a helicopter to scatter potassium (potassium fertilizer) and nitrogen (nitrogen fertilizer) from above (Fig. 12.6).

The aim of the potassium fertilization on the leaves was to inhibit the migration of radiocesium into the wood inside the trees. We found that the potassium did indeed inhibit the migration of cesium into the wood, but only to a limited degree.

We supplied nitrogen fertilizer on the leaves to promote the incorporation of cesium by the entire tree. As a result, the incorporation did increase, with as much as twice the amount of cesium absorbed in some cases. When trees were felled after scattering nitrogen fertilizer on the leaves, we verified that the trees had absorbed cesium from the soil, resulting in decontamination of the forest. This method could be used when contaminated trees are felled and cleared.

An alumnus of the Tokyo University of Agriculture (a graduate of the Department of Forestry) owned some forestland in Minamisoma's Ogai district, and we were able to use this forestland for testing purposes (Fig. 12.7). In 2012 we planted a variety of fast-growing trees there and took measurements to verify whether the planted tree species absorb radionuclides (Fig. 12.8). The results indicated that the willow and chinaberry trees absorbed radiocesium steadily (Fig. 12.9).

Fig. 12.6 A helicopter is scattering potassium and nitrogen

Fig. 12.7 Experimental forest field, given by Y. Takeyama, with a student in Minamisoma

Fig. 12.8 Students working at the forest field in Minamisoma

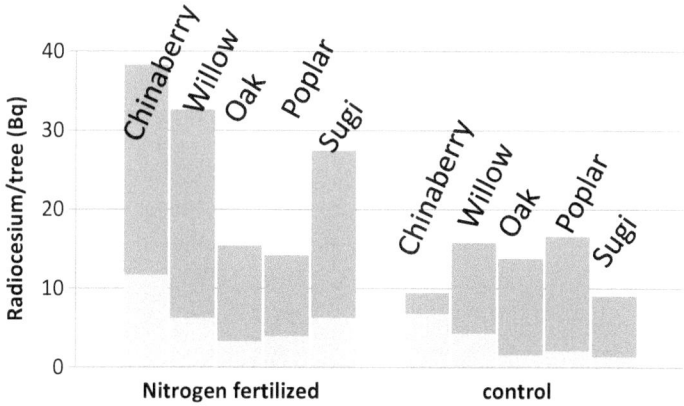

Fig. 12.9 Amount of radiocesium uptake for each tree. *Dark bars* show content of radiocesium in leaf and branch; *light bars* show content in stem

Fig. 12.10 Phytoremediation strategy in the forest field of Minamisoma by using chinaberry seedlings

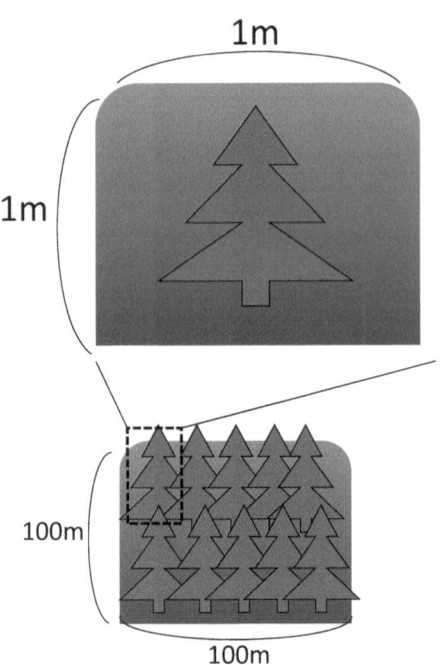

To perform a simple calculation, about 2.5 billion becquerels (Bq) of radiocesium has accumulated in each hectare of this forest. As the half-life of cesium-137 is 30.17 years, it would take about 660 years for the radiation in the soil to fall below 100 Bq/kg if the forest in Ogai was left as it is without doing anything. If willow and chinaberry saplings were planted at intervals of 1 m and only the parts above ground were cut off every few years, the time taken to bring down the cesium-137 level would be shortened to 73 years (Fig. 12.10). If nitrogenous fertilizer was added simultaneously, the time taken could be further shortened to 54 years. These fast-growing trees regenerate by sprouting, so even if their stems are cut off, they always sprout again in the spring. We are therefore considering the possibility of harvesting such trees above the ground and burning them to generate energy, rather than planting new forests.

12.4   Compensation for Forestry Operators

Forestry operators are required to take their timber to market to receive compensation from Tokyo Electric Power Company (TEPCO) for damage caused by radioactive contamination.

Felled timber is transported to the timber logistics centers in the cities of Ishinomaki or Iwaki by the forestry association or companies in the timber industry

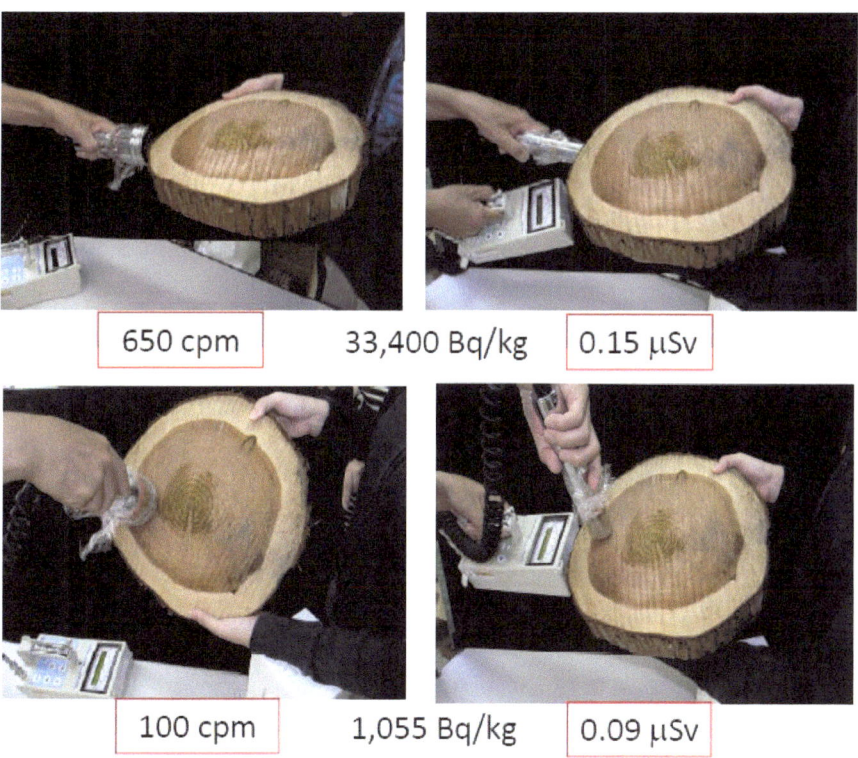

Fig. 12.11 Relationship between cpm, Bq, and μSv in wood

in Fukushima Prefecture upon request from forestry operators. If the price at which the timber sells is lower than the current market price or previous selling prices for the same timber, TEPCO pays the difference as compensation for reputational damage. This process requires submission of a large number of forms and supporting documents. In addition, at the timber logistics center in Iwaki, lumber has to be checked using a Geiger-Muller (GM) survey meter to verify that it meets the Fukushima prefectural forestry association's self-imposed limit of 1,000 cpm or less for surface radiation on lumber. The relationship between counts per minute (cpm), bequerels (Bq), and micro-sieverts (μSv) is shown in the wood (Fig. 12.11) and the wood cubics (Fig. 12.12).

If timber industry companies in other prefectures are requested to transport the timber, the felled timber is loaded onto trucks and transported to logistics centers, retailers, or large consumers in other prefectures, where it is sold. Although it is not known how much radiocesium is infiltrated inside the timber, this is not in violation of the law because the Forestry Agency declared the timber in Fukushima Prefecture to be safe.

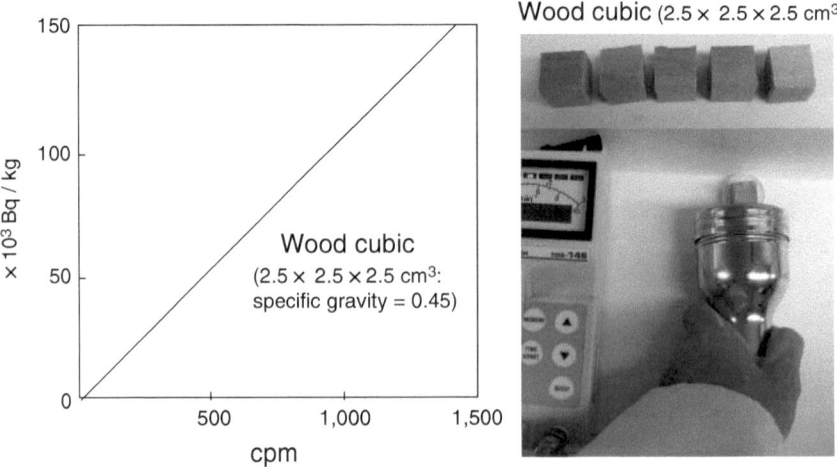

Fig. 12.12 Relationship between cpm and Bq in wood cubic

12.5 Issues Affecting the Wider Society

When traveling around the farming villages, it becomes apparent that people are using firewood (Fig. 12.13). When wood containing cesium is burned, fly ash (particulate matter) is generated, and this results in internal radiation exposure. Unfortunately, the reality is that people in these communities are not well informed about health and safety issues relating to radioactive cesium.

Meanwhile, in the forestry industry a sense of despair prevails over the contamination of forests by radionuclides. Up until the nuclear disaster, forests felled for timber were always replaced through afforestation. After the disaster, there was an increase in the number of forests abandoned without replanting (Fig. 12.14). This is particularly true in the forests in Minamisoma, and it reflects this sense of despair.

In August 2012, the Forestry Agency declared the timber in Fukushima Prefecture to be safe. This decision was based on the radioactivity measurements of 12 types of wood from tree trunks sampled from the prefecture in March 2012. The xylem of the Japanese red pine in Ohara, part of Minamisoma's Haramachi district, gave the highest reading at 497 Bq/kg. The Forestry Agency assessed that the impact on the human body would be almost nonexistent at this level of radioactivity. However, we ourselves also took measurements of timber samples from Ohara, and although some were below this level, we detected a higher level of radioactivity (1,000 Bq/kg or more) in many of the samples. Yet, despite this, no device capable of measuring radiation levels inside the timber (as opposed to on the surface) has been installed at

Fig. 12.13 Firewood used for heating bathing water and for cooking from the forests in Minamisoma on November 16, 2012

Fig. 12.14 Two forest fields with replanted trees (before the accident) and with no trees (after the accident) in Minamisoma

Fig. 12.15 Logging in the forest of Minamisoma on November 6, 2012

the logistics centers. Fukushima Prefecture needs a safe and reliable system capable of stopping the distribution of timber with a high radioactivity level.

Timber produced in Fukushima is being distributed throughout the country on the basis of the Forestry Agency's safety declaration. Visitors to the forests in Minamisoma may come across heavy machinery being used for logging (Fig. 12.15). This logged timber is being distributed throughout the country without any idea of how much radioactive cesium it contains.

Trees have to be planted in the forest, then grown for 60 to 80 years before they can be harvested and distributed as timber products. Because the forests are now contaminated by radiation, trees planted in the future may absorb large amounts of cesium-137. However, compensation for radioactivity in timber cannot be paid on an annual basis in the same way as it is for agricultural produce, so it remains doubtful how long payment of compensation will continue.

Furthermore, contamination of forests by radiation is not covered under the gS30 forest insurance scheme (Fig. 12.16). The national insurance scheme does, however, cover forest fires. After the Chernobyl nuclear disaster, many of the forests in the vicinity were set on fire deliberately to claim an insurance payout. When this happened, fly ash containing radioactive cesium was scattered, and even now people still seem to be living in fear of its effects. The Japanese are not likely to exacerbate the problems in Fukushima's forests by setting them on fire. However, they are at a loss over how to reconcile themselves to the situation.

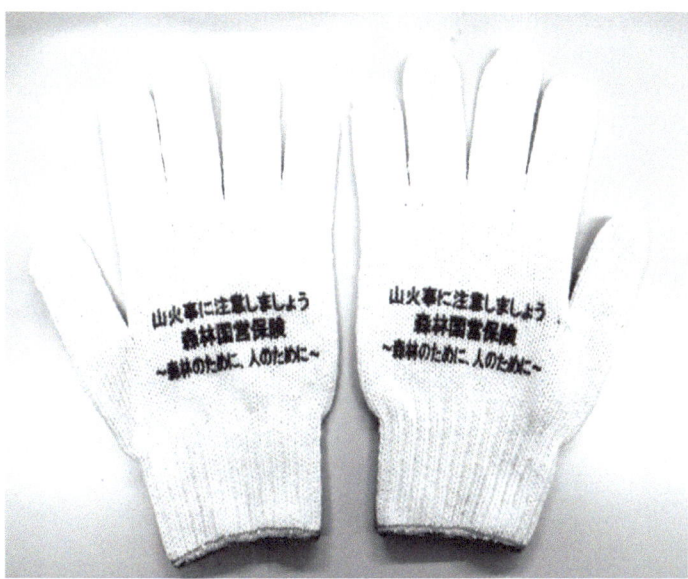

Fig. 12.16   Forest insurance does not cover the contamination of trees

Fig. 12.17   The forests have become enormous unnatural repositories of ^{137}Cs

The forests of Fukushima absorbed the radioiodine and radiocesium scattered by TEPCO's Fukushima Daiichi nuclear plant. Instead of lamenting this fact, we should be grateful that the radionuclides were largely absorbed by the trees, and this is arguably what saved the residents of Fukushima from worse contamination. These trees absorbed a large quantity of cesium-137 with a half-life of 30.17 years (Fig. 12.17), and as I pursue my research, I, for one, view that as a blessing.

Chapter 13
Nuclear Radiation Levels in the Forest at Minamisoma, Fukushima Prefecture

Iwao Uehara, Tomoko Seyama, Fumio Eguchi, Ryuichi Tachibana, Yukito Nakamura, and Hiroya Ohbayashi

Abstract The Great East Japan Earthquake occurred on March 11, 2011, and 4 days later, the Tokyo Electric Power Company (TEPCO) Fukushima No. 1 Nuclear Power Plant accident happened. The accident caused serious nuclear pollution damage for the Fukushima area, and it was reported that the forest area had received especially severe damage. However, its present situation has not been studied yet, nor has reforestation been planned. Therefore, we surveyed the amount of nuclear radiation at the forest of Minamisoma City where the amount of radiation has been reported as extremely high. We set several survey plots in the forest and surveyed the radiation amount of leaves, branches, wood (bark and stem), soil, litter interception, and irrigation water. The surveying results showed nuclear pollution was not spread equally in the Minamisoma forest, but in several "hot spots," that some litter interception indicated high radiation amounts, the extraction rates from bark to xylem were different between conifers and deciduous trees and between standing living trees and mushroom bed logs, and radioactive cesium was not detected in transpiration water.

Keywords Nuclear radiation • Minamisoma City • Fukushima • Forest

13.1 Introduction

The Great East Japan Earthquake occurred on March 11, 2011. It was followed during the next 4 days by the Tokyo Electric Power Company (TEPCO) Fukushima Daiichi Nuclear Power Station accident. The accident caused serious nuclear pollution damage in the Fukushima area, and it has been reported that forested areas sustained especially severe radiation damage (Uehara et al. 2014; Kaneko et al. 2013; Nonaka et al. 2012; Schreurs and Yoshida 2013; Takeuchi 2011). However, no

I. Uehara (✉) • T. Seyama • F. Eguchi • R. Tachibana • Y. Nakamura • H. Ohbayashi
Department of Forest Science, Tokyo University Agriculture,
1-1-1 Sakuragaoka, Setagaya-ku, Tokyo 156-8502, Japan
e-mail: i1uehara@nodai.ac.jp

© The Author(s) 2015
T. Monma et al. (eds.), *Agricultural and Forestry Reconstruction After the Great East Japan Earthquake*, DOI 10.1007/978-4-431-55558-2_13

definitive research has yet been undertaken into the consequences of the accident, and no plans exist for rehabilitating the forest. This study therefore surveyed the nuclear radiation levels in the forest at the city of Minamisoma, where radiation levels were reported to be extremely high (Ministry of Education, Culture, Sports, Science and Technology 2011).

13.2   Methods

We identified six survey areas in the Minamisoma forest (Fig. 13.1) and surveyed the radioactive cesium levels of leaves, branches, trunks (bark and xylem), soil, litter layer, and irrigation water. We also surveyed transpiration water from the leaves of trees. We used a radiation surveying system and survey meter from Hitachi Aloka Medical and a germanium detector system from CANBERRA Industries. The survey lasted from April 2012 to July 2013.

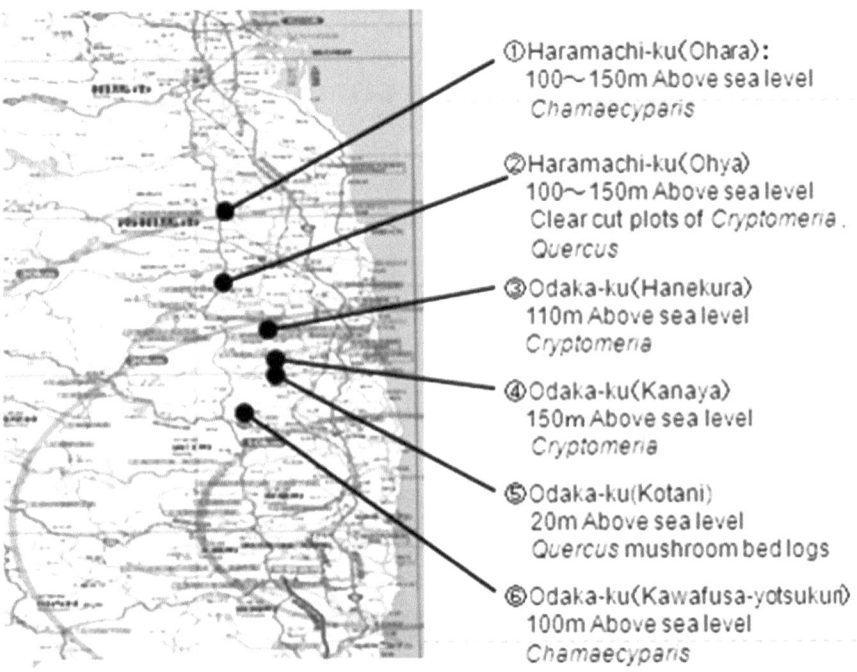

Fig. 13.1 Survey areas in Minamisoma

13.3   Results

13.3.1   Forest Soil

Radiation levels found in the soil around conifers (*Chamaecyparis* and *Cryptomeria*) were high, whereas levels around deciduous *Quercus* were lower. The radiation levels in clear-cut plots were lower than in the stands (Table 13.1).

13.3.2   Litter Layer

Some areas indicated extremely high levels of radiation. The data for Tetsuzan Dam in Haramachi and the *Chamaecyparis* stand were especially high (Table 13.2). In terms of geographic features, the former is in a valley and the latter is in a flat field without barrier materials (Figs. 13.2 and 13.3). It has already been reported that *Cryptomeria* and *Pinus* are sensitive to radiation, and the litter layer is where radioactive cesium is easily accumulated (Yoshida 2012). In addition, it was reported in the same work that evergreen trees absorb more radioactive cesium than deciduous trees. Our survey data corroborated these findings, indicating that radiation levels in evergreen conifer wood were higher than in deciduous wood.

Table 13.1 Radioactive cesium (Cs) detected at 5-cm depth in forest soil (Bq/kg)

Area	Altitude (m)	^{134}Cs	^{137}Cs	Total Cs
Haramachi-ku	135	13,600	19,100	32,700
Chamaecyparis stand (25 years)				
Haramachi-ku	120	5,250	7,460	12,710
Chamaecyparis stand (25 years)				
Haramachi-ku	20	3,220	5,550	8,770
Quercus acutissima stand				
Haramachi-ku	130	1,380	1,990	3,360
Clear-cut plot of *Quercus serrata*				
Haramachi-ku	120	78	177	255
Clear-cut plot of *Cryptomeria*				
Haramachi-ku	0	68	143	211
Minamiebi seaside				
Odaka-ku	20	14,700	22,300	37,000
Cryptomeria stand (40 years)				

Table 13.2 Radioactive cesium detected in litter layer (Bq/kg)

Area	Altitude (m)	134Cs	137Cs	Total Cs
Haramachi-ku	280	78,300	143,000	221,300
Tetsusan Dam				
Haramachi-ku	120	14,180	32,400	46,580
Cryptomeria stand (40 years)				
Haramachi-ku	80	5,340	10,800	16,140
City museum: *Abies densiflora* stand				
Haramachi-ku	120	3,830	6,680	10,510
Chamaecyparis stand (20 years)				
Odaka-ku	50	113,000	177,000	290,000
Chamaecyparis stand (15 years)				
Odaka-ku	120	53,200	93,600	146,800
Cryptomeria stand (40 years)				
Odaka-ku	100	11,300	19,200	30,500
Cryptomeria stand (40 years)				
Odaka-ku	120	7,380	12,300	19,680
Meeting house				
Odaka-ku	110	697	1,100	1,797
Horse Park: *Prunus, Zelkoba*				

Fig. 13.2 A 15-year *Chamaecyparis* stand in Odaka-ku

Fig. 13.3 Air radiation is
11–13 µSv at 15-year
Chamaecyparis stand in
Odaka-ku

13.3.3   Branches and Leaves

The levels of radioactive cesium varied considerably, but the levels in conifers and
evergreen trees were higher than in deciduous trees (Table 13.3).

13.3.4   Bark and Wood of Standing Trees

The levels of radiation in conifers were higher than in deciduous trees (Table 13.4).

13.3.5   Quercus serrata *Mushroom Bed Logs Outdoors*

Fukushima Prefecture, and the city of Minamisoma in particular, are key locations
for production of mushroom bed logs and are home to many mushroom farmers.
After the nuclear accident, however, many bed logs were abandoned outdoors
(Fig. 13.4). The levels of radiation found in these bed logs varied greatly, but the
levels found in xylem were approximately around 60 % of the levels found in bark
(Table 13.5).

Table 13.3 Radioactive cesium detected in branches and leaves (Bq/kg)

Area	Altitude (m)	^{134}Cs	^{137}Cs	Total Cs
Haramachi-ku	120	1,930	3,280	5,210
Cryptomeria (40 years)				
Haramachi-ku	120	2,130	3,010	5,140
Cephalotaxus harringtonia				
Haramachi-ku	120	1,080	2,450	3,530
Quercus myrsinifolia				
Haramachi-ku	120	1,270	1,690	2,960
Carpinus tschonoskii				
Haramachi-ku	120	1,030	1,560	2,590
Zanthoxylum piperitum				
Haramachi-ku	120	337	814	1,151
Orixa japonica				
Haramachi-ku	120	ND	171	171
Padus grayana				
Odaka-ku	50	14,500	21,500	36,000
Chamaecyparis (15 years)				
Odaka-ku	120	8,030	19,000	27,030
Cryptomeria (40 years)				
Odaka-ku	50	4,180	6,030	10,210
Callicarpa japonica				
Odaka-ku	50	1,984	3,419	5,403
Acer palmatum				

Table 13.4 Radioactive cesium detected in the bark and wood of standing trees (Bq/kg)

	Altitude (m)	Bark			Xylem			
		^{134}Cs	^{137}Cs	Total Cs	^{134}Cs	^{137}Cs	Total Cs	Xylem/bark
Haramachi-ku	60	2,187	2,275	4,462	1,358	1,654	3,012	67.5 %
Quercus serrata								
Odaka-ku	20	233	249	482	144	150	294	61.0 %
Quercus serrata no. 1								
Odaka-ku	20	362	341	703	157	184	341	48.5 %
Quercus serrata no. 2								
Odaka-ku	20	719	1,213	1,932	4,258	762	5,020	63.1 %
Quercus serrata no. 3								
Odaka-ku	50	13,418	15,418	28,836	7,619	9,418	17,037	59.0 %
Quercus serrata								

Fig. 13.4 Abandoned *Quercus serrata* mushroom bed logs in Odaka-ku

Table 13.5 Radioactive cesium detected in *Quercus serrata* mushroom bed logs outdoors (Bq/kg)

	Altitude (m)	Bark			Xylem			
		^{134}Cs	^{137}Cs	Total Cs	^{134}Cs	^{137}Cs	Total Cs	Xylem/bark
Haramachi-ku	60	2,187	2,275	4,462	1,358	1,654	3,012	67.5 %
Quercus serrata								
Odaka-ku	20	233	249	482	144	150	294	61.0 %
Quercus serrata no. 1								
Odaka-ku	20	362	341	703	157	184	341	48.5 %
Quercus serrata no. 2								
Odaka-ku	20	719	1,213	1,932	4,258	762	5,020	63.1 %
Quercus serrata no. 3								
Odaka-ku	50	13,418	15,418	28,836	7,619	9,418	17,037	59.0 %
Quercus serrata								

13.3.6 Herbaceous Vegetation and Sprouts in Clear-Cut Plots

We also studied herbaceous plants and sprouts that germinated in the spring of 2012. With the exception of *Lamiaceae*, these data show lower radiation levels than those for trees (Table 13.6).

Table 13.6 Radioactive cesium detected in herbaceous vegetation, flowers, and sprouts in clear-cut plots (Bq/kg)

	^{134}Cs	^{137}Cs	Total Cs
Follopia japonica	ND	ND	ND
Macleoya cordata	ND	ND	ND
Lamiaceae	3,320	4,720	8,040
Male flower of *Castanea crenata*	306	553	859
Sprout of *Quercus serrata* no. 1	663	1,000	1,663
Sprout of *Quercus serrata* no. 2	266	281	547
Sprout of *Quercus serrata* no. 3	257	295	552

Fig. 13.5 Trapping transpiration water from the leaves of young *Cryptomeria japonica* using plastic bags in June 2013

13.3.7 Transpiration Water and Irrigation Water

We also surveyed transpiration water from the leaves of trees (Fig. 13.5). The leaves of 8-year-old and 40-year-old *Cryptomeria* trees were wrapped in plastic bags (27 cm^2) for 24 h on June 13, 2013. Samples of 20 ml were analyzed, but radioactive cesium was not detected. This result suggests that the possibility of secondary radioactive contamination from living standing trees may be low.

We also surveyed irrigation water at some locations in the Haramachi and Odaka areas, but radioactive cesium was not detected.

13.4 Discussion

The foregoing results suggest these points.

1. Levels of nuclear radiation in Minamisoma were influenced mainly by the city's location to the northwest of the Fukushima Daiichi Nuclear Power Station and the prevailing wind direction on March 14, 2011.
2. High levels of radiation were distributed unevenly over valleys or fields without barriers.
3. In general, the nuclear pollution was not evenly distributed, but was in the form of "hot spots."
4. The litter layer in some areas showed extremely high levels of radiation.
5. Abandoned forests and multi-layered forests indicated high levels of radiation; levels were lower in clear-cut plots.
6. The extraction rates from bark to xylem were different between conifers and deciduous trees. They were also different between standing living trees and mushroom bed logs (*Quercus serrata*).
7. Levels of radiation found in mushroom bed logs abandoned in the outdoors varied greatly.
8. Radiation levels of herbaceous plants and *Quercus serrata* sprouts that germinated in the spring of 2012 were clearly lower than those of trees in stands.
9. No radioactive cesium was detected in transpiration water, suggesting that the probability of secondary radioactive contamination from living standing trees may be low.

Some attempts need to be made to decontaminate forest areas in Fukushima Prefecture, but the prefecture's forests are extensive and comprise geographically complex terrain, so it is not practically possible to decontaminate them all. However, we propose creating small clear-cut examination plots (Fig. 13.6) in locations where soil erosion cannot occur within the forests. Our survey data showed that the radiation levels of herbaceous vegetation and *Quercus serrata* sprouts germinated in the clear-cut plot in spring 2012 were clearly lower than those of trees in stands. One practical method, therefore, would be to promote growth of the newly germinated plants and sprouts in such plots while continuing to survey radioactive cesium levels.

13.5 Conclusion

The results of this study showed that there were still forests in which radioactive cesium levels were high. The half-life of cesium is more than 30 years, and forest regeneration also takes a long time. We should therefore continue to collect data on radioactive cesium and the dynamics of vegetative regeneration.

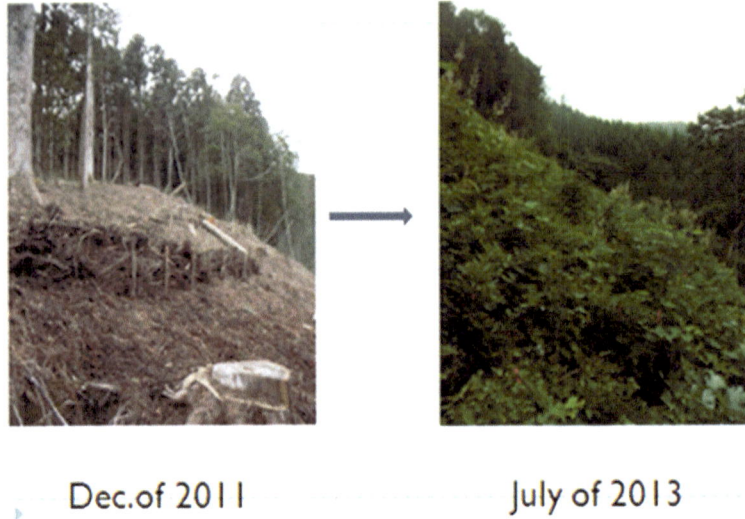

Dec.of 2011   July of 2013

Fig. 13.6 Clear-cut plot within a *Cryptomeria japonica* stand in Haramachi, Minamisoma

Acknowledgments We wish to express our appreciation to Eihachi Horiuchi, chief of Soma's regional forestry cooperative, who helped with our survey, and Yoichi Takeyama, who provided us with experimental forest plots and guided us around Minamisoma's forests.

References

Kaneko S et al (2013) The change of air dose rate in the forest contaminated with radioactive materials by the accident at Tokyo Electric Power Company's Fukushima Daiichi Nuclear Power Plant. Kanto J For Res 64(2):37–40 (in Japanese)

Ministry of Education, Culture, Sports, Science and Technology (2011) Total cesium data on the ground surface. http://ramap.jmc.or.jp/map/eng/map.html (in Japanese)

Nonaka M, et al (2012) The dynamics of radioactive cesium in the wood. In: The 123th Japanese Forest Society annual meeting transaction (CD-R) (in Japanese)

Schreurs M, Yoshida F (2013) Fukushima: a political economic analysis of a nuclear disaster. Hokkaido University Press, Sapporo

Takeuchi K (2011) There has been still big unconvergence problems at Fukushima nuclear plant: it has come to show same condition of Chernobyl. Green Power 2011:24–25 (in Japanese)

Uehara I, et al (2014) Continued radiation research at Minamisoma City forest, Fukushima, from 2012 to 2014. In: 4th Kanto Forest Research meeting transactions (in Japanese)

Yoshida S (2012) The movement of radioactive fallout in the forest ecosystem: estimated condition and problems considered by past studies. For Sci 65:31–33 (in Japanese)

Chapter 14
Radioactive Contamination of Ostriches in a Potentially Permanent Evacuation Zone

Hiroshi Ogawa, Hidehiko Uchiyama, Koji Masuda, Takeshi Sasaki, Tadao Watanabe, Toshiaki Tomizawa, and Schu Kawashima

Abstract We measured the concentrations of radioactive cesium (^{134}Cs, ^{136}Cs, and ^{137}Cs) in muscles and internal organs of ostriches that had been contaminated by radioactive materials from the Fukushima Daiichi Nuclear Power Station. Although the ostriches were given uncontaminated feed and water throughout the study, radionuclide could not be eliminated completely as the birds consumed soil and plants in the enclosures. However, we found that the amount of radioactive cesium in the ostriches' bodies declined in the first 287 days after capture. This result suggests that keeping the birds in a radiation-free environment is effective in reducing the amount of radioactive cesium in their bodies.

Keywords Fukushima nuclear disaster • Ostrich • Radioactive cesium • Internal organs • Muscles

14.1 Background and Purpose of Research

The Great East Japan Earthquake of March 11, 2011, triggered an accident at the Fukushima Daiichi Nuclear Power Station, releasing a large amount of radioactive material into the environment. Because of the resulting evacuation, 30 ostriches were abandoned at an ostrich farm approximately 7 km from the power station (Fig. 14.1). Around 4 months after the accident, half the ostriches had died and some had escaped from the farm. It is believed that, following the earthquake and associated disasters, the abandoned ostriches survived by eating mainly

H. Ogawa (✉) • H. Uchiyama • K. Masuda • T. Sasaki • S. Kawashima
Department of Human and Animal-Plant Relationships, Tokyo University of Agriculture, Funako, Atsugi, Kanagawa 243-0034, Japan
e-mail: ogawah@nodai.ac.jp

T. Watanabe
The Research Institute of Evolutionary Biology, Kamiyoga, Setagaya, Tokyo 158-0098, Japan

T. Tomizawa
Strauss Co., Ltd., Yoshikawa, Saitama 342-0055, Japan

© The Author(s) 2015   203
T. Monma et al. (eds.), *Agricultural and Forestry Reconstruction After the Great East Japan Earthquake*, DOI 10.1007/978-4-431-55558-2_14

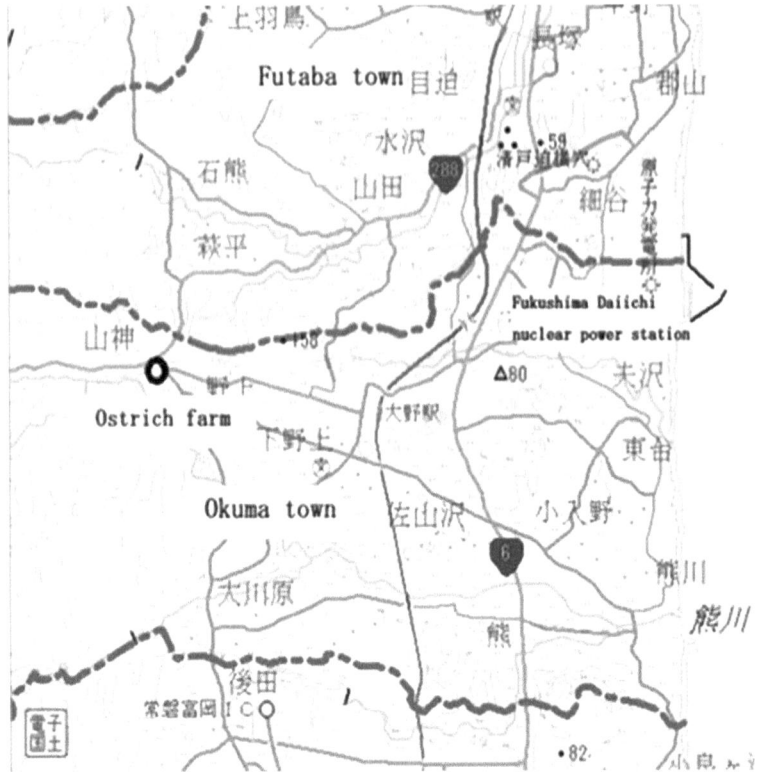

上羽鳥
Futaba town 目迫
郡山
水沢
石熊   山田
萩平
細谷

Fukushima Daiichi
山神   nuclear power station
△80   夫沢
Ostrich farm
大野駅
Okuma town 左山沢   小入野
大川原
熊川
熊川
後田
常磐富岡IC
・82・

Fig. 14.1 Survey site

radiation-contaminated plants and insects and drinking standing water such as rain-water puddles. We had reason to suspect, therefore, that the radionuclide scattered over the area had accumulated in their bodies. Other livestock left behind were in most cases not permitted to be taken out of the 20-km exclusion zone and as a rule were euthanized. Under these circumstances, we attempted to record the extent to which the escaped ostriches had been contaminated, as well as identify how the concentrations of radiation inside their bodies changed when they were provided with uncontaminated feed. We believed such a study could help elucidate the extent to which animals in general had been internally contaminated, the effects on their health caused by the contamination, and the process by which it could be decreased or eliminated. We also believed the study could help identify some key principles to be applied when resuming agricultural production in the affected areas.

14.2 Materials and Methods

14.2.1 The Ostriches Examined

The study was conducted on five male ostriches and one female ostrich (African Black; Fig. 14.2) that had been recaptured after escaping from the farm in the town of Okuma in Futaba County, Fukushima Prefecture.

14.2.2 Handling and Feeding Methods

The ostriches were kept in 180-m^2 enclosures with a 32-m^2 roofed area (Fig. 14.3), either on their own, or with a maximum of one male and two females per enclosure (one female ostrich that died during the study was not included in the results). Commercial dog food and clean well water were supplied as feed and drinking water. Food and drinking water were replenished once per week. However, in March 2012 they were replenished once per fortnight.

Fig. 14.2 Two of the ostriches examined

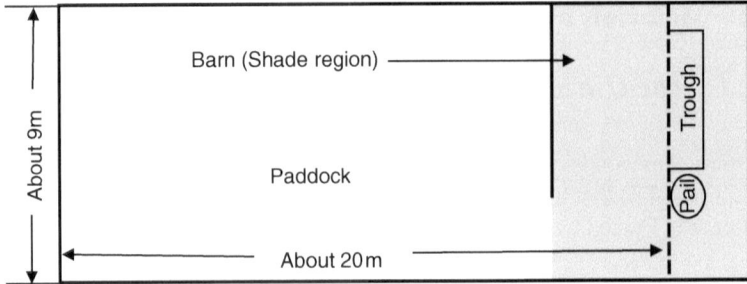

Fig. 14.3 Overview of enclosure

Table 14.1 Dates of capture and examination

Recapture day	Sex	Date examined	Days from recapture to examination	Age in months
January 12, 2012	Male	March 27, 2012	74	79
January 13, 2012	Male	February 16, 2012	34	66
January 13, 2012	Male	October 26, 2012	287	180
January 13, 2012	Female	February 8, 2013	392	60–84
January 13, 2012	Male	February 21, 2013	405	60–84
May 29, 2012	Male	May 25, 2013	361	Unknown

14.2.3 Study Method

The number of days between capturing and examining the ostriches ranged from 34 to 405 days (Table 14.1). Before being examined, the ostriches were restrained and euthanized via injection of anesthetic (somnopentyl) into the neck vein.

14.2.4 Body Measurement and Collection of Samples

After taking body measurements of the dead ostriches, including total body length, maximum wing length, wingspan, tail length, exposed beak length, total beak length, total head length, beak width, toe length, toe span, claw length, and tarsus length, the birds were dissected and samples collected.

14.2.5 Radiation Measurement

We took samples for radiation measurement purposes from the chest muscle, subcutaneous chest fat, lower and upper leg muscles, brain, thyroid, liver, spleen, crop, gizzard, stomach content, rectal content, pancreas, heart, lungs, kidneys, adrenal gland, testes or ovaries, egg yolk, egg white, and eggshell. Plants, soil, and drinking water were also collected as environmental samples. We measured radionuclide by placing samples in a 100-ml vessel and using a germanium semiconductor detector to identify gamma-ray nuclides and the concentration of radioactive material (Bq/kg) via the gamma-ray spectrometer method.

14.3   Research Findings

14.3.1   Ambient Dose

The ambient radiation doses measured during the study were as follows:

March 16, 2012: 4.6 μSv/h
March 27, 2012: 4.0 μSv/h
October 24, 2012: 3.4 μSv/h
February 8, 2013: 2.3 μSv/h
February 21, 2013: 2.4 μSv/h
May 25, 2013: 2.0 μSv/h

Thus, dosage per hour gradually declined with time.

14.3.2   Body Measurements

The ostriches examined had an average total body length of 232 cm, maximum wing length of 123 cm, wingspan of 231 cm, tail length of 30 cm, exposed beak length of 6 cm, total beak length of 9 cm, total head length of 19 cm, beak width of 7 cm, toe length of 19 cm, toe span of 24 cm, claw length of 4 cm, and tarsus length of 45 cm. Comparison of measurements such as beak, head, toe, and tarsus length did not reveal any significant difference in physique between males and females. However, the male population surveyed included physically small birds, and variance in growth and physical condition between specimens was observed.

14.3.3   Pathological Findings Relating to Internal Organs

The state of the digestive organs (stomach, gizzard, intestines), trunk muscle, bone (cranium), respiratory organs (trachea, lungs), circulatory organs (blood vessels, heart), and various tissues (thyroid, etc.) showed the specimens to be in relatively good health given the fact that they had been wandering outdoors for some time.

14.3.4   Radionuclide in Test Samples

Radioactive cesium in the ostriches' living environment was 2,300–75,000 Bq/kg in the soil and 1,200–4,400 Bq/kg in nearby plants. No radiation was detected in drinking water, despite adulteration with rainwater.

Radioactive cesium in stomach content ranged from 3,100 to 20,000 Bq/kg, despite the fact that the ostriches were given uncontaminated feed, likely from consumption of soil or plants within or outside the enclosure, rather than the clean feed administered. The highest concentration of radioactive cesium was found in rectal content, ranging from 11,000 to 65,000 Bq/kg, likely the result of concentration in the digestive organs.

In general, however, levels of radioactive cesium in the bodies of the ostriches tended to be low in the internal organs and higher in the skeletal muscles. In contrast to concentrations of 1,100–6,700 Bq/kg in the gizzard and heart, which are composed almost completely of internal muscle tissue, cesium was highly concentrated in the skeletal chest and leg muscles at 2,300–27,000 Bq/kg. Concentration in other internal organs ranged from 500 to 6,700 Bq/kg. Radioactive cesium contained in eggs was 500–1,900 Bq/kg in the whites, 200–740 Bq/kg in the yolks, and 70–220 Bq/kg in the shells. The concentration in the whites was therefore 2.5 to 2.7 times that in the yolks, likely related to the high proportion of fat in the yolk. A minute amount of potassium-40 was also detected in the kidneys, heart, testes, and thyroid.

Although the ostriches were given uncontaminated feed and water throughout the study, radionuclides could not be eliminated completely as the birds consumed soil and plants in the enclosures. However, we found that the amount of radioactive cesium in the ostriches' bodies declined in the first 287 days after capture, before leveling off with slight fluctuations, suggesting that keeping the birds in a radiation-free environment is effective in reducing the amount of radioactive cesium in their bodies.

14.4   Conclusion

Although radioactive cesium emitted by the Fukushima nuclear accident was detected in the bodies of ostriches abandoned in the town of Okuma, the level declined when the birds were given clean feed and water. As outlined in the Bergonié–Tribondeau law (Bergonié and Tribondeau 1906), it is known that the effect of radiation on biological tissue is greater in cells that undergo frequent cell division, cells that will undergo numerous cell divisions in future, and undifferentiated cells[1]. Although much is still unknown regarding long-term exposure to lC radiation in humans, studies on the impact of the Chernobyl nuclear accident indicate effects such as an increase in tumors in children, particularly in the occurrence of thyroid cancer, and a negative impact on pregnant women and the development of fetuses (Investigation Committee for Chernobyl nuclear power plant accident, etc. of the House of Representatives 2011). To eliminate the effect of radionuclides, it is important that the living environment is kept as free of radiation as possible and that uncontaminated food is consumed on an ongoing basis.

Acknowledgments   This research received funding from the Society for Livestock Studies. We thank Dr. Daishiro Yamagiwa, who provided tremendous assistance for this study.

References

Bergonié J, Tribondeau L (1906) De Quelques Résultats de la Radiotherapie et Essai de Fixation d'une Technique Rationnelle. C R Seances Acad Sci 143:983–985
Investigation committee for Chernobyl Nuclear Power Plant Accident, etc. of the House of Representatives (2011) A report by investigation committee for Chernobyl nuclear power plant accident, etc. of the House of Representatives. http://www.shugiin.go.jp/itdb_annai.nsf/html/statics/shiryo/201110cherno.htm, 2 Sept 2014

Chapter 15
Radioactive Contamination in Some Arthropod Species in Fukushima

Tarô Adati and Sota Tanaka

Abstract To clarify the extent of radioactive contamination in a broad environment of Fukusima, in and around farming lands and residential areas, the amount of radionuclides in the Japanese grasshopper, *Oxya yezoensis*, the Emma field cricket, *Teleogryllus emma*, the wasp spider, *Argiope bruennichi*, and the Jorô spider, *Nephila clavata*, was investigated. Radioactive cesium was detected in all arthropods collected from survey sites in Fukushima. The highest radioactive cesium (^{134}Cs and ^{137}Cs) concentration of approximately 4.7×10^2 Bq/kg (wet weight) was detected from the grasshopper. The amount of radionuclides tended to rise in agreement with the space radiation dose rates in the survey sites. The spiders, which are classified at higher trophic levels on the food chain, therefore demonstrated correspondingly higher levels of cesium concentration, suggesting that bioaccumulation of radioactive cesium was occurring.

Keywords Insects • Spiders • Environmental indicator • Food chain • Bioaccumulation

15.1 Introduction

15.1.1 Radiation and the Ecosystem

Although 2 years have passed since the Fukushima Daiichi Nuclear Power Plant (FDNPP) accident, the recovery of agriculture in many areas of Fukushima Prefecture remains at a standstill. A major factor behind this hiatus is environmental

T. Adati (✉)
Department of International Agricultural Development, Tokyo University of Agriculture, 1-1-1 Sakuragaoka, Setagaya-ku, Tokyo 156-8502, Japan

S. Tanaka
Department of International Agricultural Development, Tokyo University of Agriculture, 1-1-1 Sakuragaoka, Setagaya-ku, Tokyo 156-8502, Japan

Research Reactor Institute, Kyoto University, Kumatori, Sennan, Osaka 590-0494, Japan

© The Author(s) 2015
T. Monma et al. (eds.), *Agricultural and Forestry Reconstruction After the Great East Japan Earthquake*, DOI 10.1007/978-4-431-55558-2_15

contamination by radionuclides. Although decontamination is progressing, mainly in rice paddies and non-paddy arable fields, shipment restrictions on some crops in certain areas remain as of July 2013, including fruit varieties such as yuzu (citron), kiwifruit, and chestnuts, fungi such as shiitake mushrooms, and wild vegetables such as bamboo shoots and bracken shoots (Forestry Agency 2013; MAFF 2013). Despite decontamination of fields and areas near residences, moreover, in some cases the ambient radiation dose rate rises again as time passes, requiring re-decontamination.

Decontamination of cropland generally involves removing the top layer of soil, including plants contaminated with radioactive fallout from the FDNPP. When determining which areas to decontaminate, focus is placed on districts in which people are living, although land usage and administrative boundaries also are factors in decisions. However, in Fukushima, which has large areas of forest-covered mountains, inhabited districts cover a comparatively insignificant area, and even if local residential zones are decontaminated, subsequent rain or wind will bring radiation down from the mountains, re-contaminating them.

The landscape of Japan's farming villages resembles a mosaic, featuring not only rice paddies and non-paddy arable fields but also waterways, levees, and ponds, as well as houses and surrounding tree groves, backed by woodland that spreads into the mountains. This setting is home to a diverse range of life, including aquatic organisms (insects, crustaceans, shellfish, amphibians, fish, etc.), land organisms (insects, spiders, reptiles, mammals, etc.), birds, and plants (crops and weeds). This farming village ecosystem is not closed: each organism is linked through its activities to the outside world, in an open, dynamic environment.

Indeed, nature itself is inherently dynamic, and accordingly current decontamination methods, which target relatively stationary elements such as soil and plants, are unable to decontaminate the wide-ranging ecosystem in its entirety. To do this, the extent of radioactive contamination in this broad environment must first be ascertained. Effective environmental indicators for this purpose are organisms that are widespread throughout the area as components of the ecosystem. Organisms suggested as candidates for this indicator role are generally arthropods such as insects.

So what effect *do* radionuclides have on arthropods? Or more simply—how resilient are arthropods against radiation?

There is a pest control method that uses artificial radiation to eliminate insect pests that cause damage to crops and livestock. This process entails directing gamma rays or electron beams at insects to either kill them directly or render them infertile via sterilization before returning them to their environment. Detailed data have been gathered on the effects of radiation on the insects targeted by this process.

For example, the melon fly, *Bactrocera cucurbitae* (Coquillett) (Diptera: Tephritidae), once ravaged plants and fruit of the melon family on the Nansei Islands to the south of the Japanese archipelago before being eradicated via sterilization. It is known that the fly's reproductive cells are destroyed upon exposure to 70 Gy gamma radiation from ^{60}Co, rendering the organism infertile (Koyama 1994). In addition, it is known that the sweet potato weevil, *Cylas formicarius* (Fabricius)

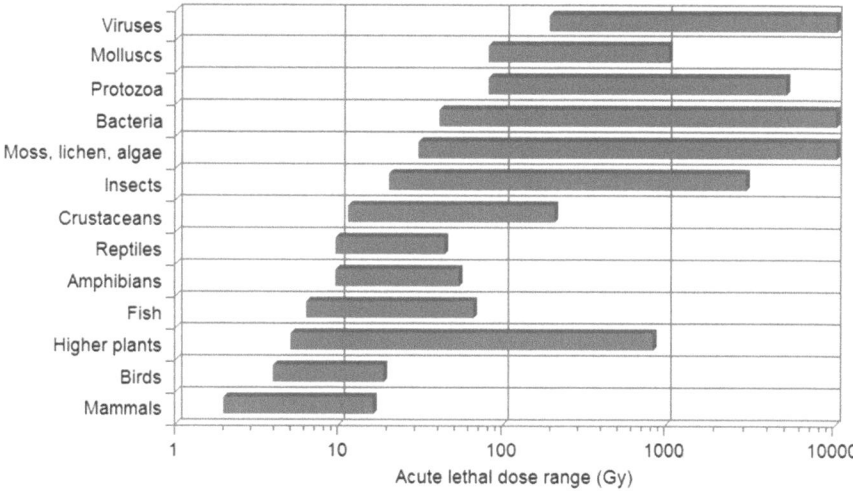

Fig. 15.1 Acute dose ranges that result in 100 % mortality in various taxonomic groups (Modified from IAEA 2006)

(Coleoptera: Brentidae) is also sterilized upon irradiation with gamma rays of 70 Gy or greater (Iwamoto et al. 1990), and an elimination program targeting the weevil is currently in progress on the Nansei Islands. In these instances, although the irradiation process sterilizes the organism, it has little to no effect on factors other than reproduction, such as survival rate or mating behavior. Gamma radiation of 70 Gy is equivalent to 70,000 mSv, a level that would instantly kill a human. As shown in Fig. 15.1, arthropods such as insects and crustaceans are relatively tolerant of radiation within the animal kingdom, whereas humans are among the most sensitive mammals and therefore among the most sensitive organisms (IAEA 2006).

However, in contrast to artificial radiation, at present little is known of the effect of environmental radionuclides on arthropods in the wild, even from the experience of the Chernobyl and FDNPP accidents. Despite the reported discovery of malformed insects in the irradiated zones after these two nuclear accidents (Hesse-Honegger and Wallimann 2008; Hiyama et al. 2012), it is not clear whether such malformations were in fact a result of radionuclides emitted by the accidents. No conclusion can be drawn because malformation in insects from unknown causes is often observed in nature.

Following the Fukushima accident, studies of the level of space radiation and the radionuclides in soil and crops have been conducted at various sites. However, reports related to arthropods are extremely limited (Fugo 2012; Hiyama et al. 2012). One area that is decidedly lacking data is internal irradiation from radionuclides. There is only one way to investigate this: by collecting arthropods from contaminated sites and measuring the amount of radionuclides accumulated inside their bodies.

15.1.2 Traveling to Fukushima

Tarô Adati, one of the authors of this chapter, was aware that the situation 1 year on from the nuclear accident was broadly as outlined in the previous section. However, he did not intend to visit Fukushima to study arthropods, because he believed this sort of study should be performed by well-organized teams of experts rather than a lone academic, who despite specializing in entomology was an amateur in the field of radiobiology. But as chance would have it, something unexpected occurred. Sota Tanaka was a fourth-year student whose graduation research project Adati happened to be in charge of supervising. Tanaka mentioned when selecting his research topic that he wanted to investigate something that would aid Tohoku's disaster zone. He asked to study the insects in Fukushima, and faced with his enthusiasm, Adati eventually came around to the idea. So that is how a young student's passion swayed a worn-out, middle-aged academic, leading to the creation of a very small research team composed of a duo with a generation gap.

We traveled from Tokyo to Fukushima on the bullet train and rented a car near the station. We were headed for the village of Iitate. The wind direction at the time of the accident meant that the area had sustained a large amount of radioactive fallout despite being located a relatively long distance from the nuclear plant. Most of the village had been declared a long-term evacuation zone, and part of its southern end had been designated a potentially permanent evacuation zone.

We visited a certain local resident who ran a farm in the village. He was a relative of a student in our department, and we had been in contact with him beforehand. Although the district in which his house and farm were located had recently been changed to a short-term evacuation zone in which the evacuation order would soon be lifted, he was not allowed to stay in his own house overnight. However, he was working part-time guarding the village administrative office, and after working night shifts he would rest at his home during the day. He told us that, despite the continuing restrictions, the space radiation dose in his area was relatively low, perhaps a result of the topography. Once the decontamination was complete, he planned to resume farming.

15.2 Surveys in Fukushima

15.2.1 Space Radiation Dose Rates in Survey Sites

The following is an overview of the study we conducted in Fukushima in 2012. We began by using a portable NaI (Tl) scintillation survey meter to measure the space radiation dose rate at two survey sites in Iitate and one survey site in the city of Soma. Comparison of the results with data measured in the city of Atsugi in Kanagawa Prefecture, about 270 km from the FDNPP, showed correspondence between ambient radiation levels and distance from the nuclear plant (Table 15.1).

Table 15.1 Space radiation dose rates and radioactive contamination in arthropods at survey sites (Sept.–Oct. 2012)

Survey site		Distance from FDNPP (km)	Median of space radiation dose rates (μSv/h)[a]	Median of radioactive cesium (^{134}Cs + ^{137}Cs) amounts in arthropod samples (Bq/kg fresh weight)[b]			
				Japanese grasshopper	Emma mole cricket	Wasp spider	Jorô spider
A	Iitate, Fukushima	40	3.74	469	156	114	310
B	Iitate, Fukushima	44	1.98	188	194	–	214
C	Sôma, Fukushima	46	1.14	3	76	6	–
D	Atsugi, Kanagawa	267	0.04	2	8	–	4

[a]Space radiation dose rates were measured 1 m above ground surface with 5 replications per site
[b]Radioactive contamination was detected in an arthropod sample containing 10–50 individuals with 4 replications. If the radioactive measurement of a sample was lower than the measurable limit, the limit value was considered as its amount.– No samples collected

Ambient radiation dose rates exceeded 1 μSv/h at each of the three sites in Fukushima Prefecture, clearly from the effect of radioactive fallout from the FDNPP accident. However, the variance recorded among the three survey sites is likely caused by a range of factors, such as topography and wind direction at the time of the accident, rather than distance from the nuclear plant alone. The highest dosage measured during the survey was at Survey Site A, a flat area of land in a long-term evacuation zone. Site B was a fallow field in a short-term evacuation zone within an area surrounded by woodland and rivers. Although Site C was a fallow field that had yet to be decontaminated, it had a relatively low dosage compared to the other survey sites, possibly because of the decontamination of nearby fields.

15.2.2 Sampling of Arthropods and Detection of Radioactive Cesium

Next, we collected arthropod samples at each survey site (Fig. 15.2) and used a Ge detector to investigate the levels of radioactive contamination. The arthropods sampled were the Japanese grasshopper, *Oxya yezoensis* Shiraki (Orthoptera: Catantopidae) (Fig. 15.3); the Emma field cricket, *Teleogryllus emma* (Ohmachi & Matsuura) (Orthoptera: Gryllidae) (Fig. 15.4); the wasp spider, *Argiope bruennichi* (Scopoli) (Araneae: Araneidae) (Fig. 15.5); and the Jorô spider, *Nephila clavata* L. Koch (Araneae: Nephilidae) (Fig. 15.6). The Japanese grasshopper is herbivorous, the Emma field cricket is omnivorous, and the wasp and Jorô spiders are predatory.

Fig. 15.2 Collecting arthropods at a survey site in Fukushima

Fig. 15.3 The Japanese grasshopper, *Oxya yezoensis*

Fig. 15.4 The Emma field cricket, *Teleogryllus emma*

Fig. 15.5 The wasp spider, *Argiope bruennichi*, feeding on a Japanese grasshopper

Fig. 15.6 The Jorô spider, *Nephila clavata*

As shown in Table 15.1, radioactive cesium was detected in all arthropods collected from survey sites A and B. At Survey Site C, radioactive cesium was detected in the Emma field cricket, but not at significant levels in the other arthropod types. The combined amount of ^{134}Cs and ^{137}Cs per kilogram (Bq/kg) of wet weight tended to increase in line with the space radiation dose rates in the survey sites.

At Site B, the predatory Jorô spider recorded the highest contamination of radioactive cesium at about $2.1 \times 10_2$ Bq/kg; levels in the herbivorous Japanese grasshopper and omnivorous Emma field cricket were lower. Arthropods at higher trophic levels on the food chain therefore demonstrated correspondingly higher levels of cesium concentration, suggesting that bioaccumulation of radioactive cesium was occurring. However, at Site A the Japanese grasshopper recorded a radioactive cesium concentration of about 470 Bq/kg, which was higher than the other arthropod types, including spiders.

15.3 Discussion

The presence of radioactive cesium in herbivores (primary consumers) such as the Japanese grasshopper can only be attributed to consumption of radioactive fallout from the the FDNPP that contaminated the grasses (producers) which serve as the species' main food source. Furthermore, if predatory spiders (secondary consumers) feed on several herbivorous or omnivorous organisms, radionuclides are likely to be concentrated in these predatory organisms, which are higher on the food chain. In fact, we observed a wasp spider feeding on a Japanese grasshopper at Survey Site A (Fig. 15.5). However, if the radioactive concentration in higher predator spiders is low, as at Site A, it is also possible that radioactive cesium does not accumulate readily in the primary consumers such as grasshoppers for some reason. During this study, for example, we immediately preserved the collected arthropods in ethanol. Thus, grasses consumed by the Japanese grasshoppers may have remained within the alimentary canal. As a result, it is possible that most of the radioactivity detected was attached to the undigested plant material and therefore concentrated in the digestive tract. Indeed, it has been reported that the concentration of radioactive cesium declines when grasshoppers are collected and left to excrete droppings for a period of time (Fugo 2012). The transfer of radionuclides from food to body tissue in arthropods therefore needs to be investigated further in future.

References

Forestry Agency (2013) The present situation in shipment restrictions of mushrooms and edible wild plants in Fukushima Prefecture (in Japanese. Title translated by present authors). http://www.rinya.maff.go.jp/j/tokuyou/kinoko/qa/seigenfukusima.html. Retrieved on October 4, 2013

Fugo H (2012) Are radionuclides accumulated in insects? Nat Conserv 526:12–13 (in Japanese. Title translated by present authors)

Hesse-Honegger C, Wallimann P (2008) Malformation of true bug (Heteroptera): a phenotype field study on the possible influence of artificial low-level radioactivity. Chem Biodivers 5:499–539

Hiyama A, Nohara C, Kinjo S, Taira W, Gima S, Tanahara A, Otaki JM (2012) The biological impacts of the Fukushima nuclear accident on the pale grass blue butterfly. Sci Rep 2:570

IAEA (International Atomic Energy Agency) (2006) Environmental consequences of the Chernobyl accident and their remediation: twenty years of experience. Report of the Chernobyl Forum Expert Group 'Environment'. IAEA, Vienna

Iwamoto J, Ito S, Mano M, Yamasaki H (1990) Gamma sterilization of sweet potato weevil, Cylas formicarius (Fabricius) (Coleoptera: Curculionidae): effects of irradiation on fertility, longevity, and mating potential. Res Bull Pl Prot Japan 26:69–72 (in Japanese with English summary)

Koyama J (1994) Overview of the studies on the eradication of the melon flies in Japan. Jpn J Appl Entomol Zool 38:219–229 (in Japanese)

MAFF (Ministry of Agriculture, Forestry and Fisheries) (2013) The fruit production: frequently asked questions (in Japanese. Title translated by present authors) http://www.maff.go.jp/j/kanbo/joho/saigai/kazyu_seisan_qa.html. Retrieved on October 4, 2013

Chapter 16
A Consumer Survey Approach to Reputation-Based Damage Affecting Agricultural Products and How to Overcome It

Puangkaew Lurhathaiopath, Shizuka Matsumoto, Makoto Hoshi, Sayaka Yamaguchi, and Toshiyuki Monma

Abstract In this chapter we elucidated the changes in consumers' feelings of safety and reassurance and the effectiveness of safety-related PR and events to support the recovery. The new knowledge presented in this chapter can be summarized as follows. According to the results, the effective ways to overcome reputation-based damage relating to agricultural products produced in irradiated areas are (a) provide accurate information on radioactive contamination (dispel ambiguity); (b) use methods that come as close as possible to testing every bag of produce, and require that agricultural products shipped to the market contain no radionuclide detectable using standard detectors (reduce importance); and (c) secure supporters by holding events to support the recovery efforts in disaster zones. It is, moreover, important to implement such initiatives on a continuous basis.

Keywords Reputation • Radioactivity testing methods • Radiation limits • PR activities

16.1 What Is Reputation-Based Damage?

On the surface it would appear that the extreme aversion to agricultural products from Fukushima Prefecture witnessed in the aftermath of the nuclear disaster has waned, but in reality the damage resulting from negative reputation has not really

P. Lurhathaiopath (✉)
Faculty of Life and Environmental Sciences, Tsukuba University,
1-1-1 Tennodai, Tsukuba City, Ibaraki Prefecture 305-8577, Japan
e-mail: Lurhathaiopath.pu.fw@u.tsukuba.ac.jp

S. Matsumoto • M. Hoshi • S. Yamaguchi • T. Monma
Department of International Biobusiness Studies, Tokyo University of Agriculture,
1-1-1 Sakuragaoka, Setagaya-ku, Tokyo 156-8502, Japan

© The Author(s) 2015
T. Monma et al. (eds.), *Agricultural and Forestry Reconstruction After the Great East Japan Earthquake*, DOI 10.1007/978-4-431-55558-2_16

221

abated. Agricultural products from Fukushima fetch only low prices within the market, and even if supermarkets sell such products, they always take the precaution of stocking alternative products from other areas. The fact remains, therefore, that even if Fukushima produce meets safety-related criteria, there are still consumers who shun it.

Although reputation-based damage is defined in a number of ways, Sekiya (2003) defines it as "economic damage due to disrupted consumption or tourism as a result of people viewing foods, products, and localities once deemed safe as being dangerous because of widespread media coverage of an incident, an accident, environmental pollution, or a disaster." The American psychologists Allport and Postman (Satou 2004) define aggregate reputation-based damage as the importance multiplied by the ambiguity, where importance is assessed in terms of the effect on risk to life whereas ambiguity is assessed in terms of the credibility of information provided.

In this chapter, we assess the current level of reputation-based damage according to consumer sentiment regarding the safety of agricultural products since the radioactive contamination. We also evaluate potential methods of overcoming this problem. We use the results of our consumer survey to report on three issues in particular: (1) whether consumer awareness regarding the safety of agricultural products has changed 2 years after the disaster, (2) whether the prefectural authority's testing of every bag of produce is effective, and whether the radiation limits and inspection methods used to ensure safety are considered credible, and (3) the benefits derived from events to support recovery, as well as PR activities promoting the safety of Fukushima's agricultural products.

16.2   Consumer Sentiment with Regard to Reputation, Radioactivity Testing Methods, and Credibility of Information Provided

16.2.1   Overview of Radioactive Content Limits and Consumer Survey

New limits for radioactive cesium content in foods were put in place in April 2012. The new criteria reduce the five food categories used previously to four, namely, general foods, infant food products, milk and dairy products, and drinking water. They are also more stringent than the previous limits. Yet, despite the fact that the Japanese limits are lower than those set by international institutions, the EU, and the United States, Japanese consumers continue to harbor concerns over food safety (Table 16.1).

In December 2011 we conducted consumer surveys of 200 visitors each at two farmers' markets, one in Kanagawa Prefecture's Ashigara region and one in Fukushima Prefecture's Aizu region. Subsequently, in 2012, we administered two more surveys, one targeting approximately 200 people who attended the "Agricultural

Table 16.1 Comparison of limits for radioactive cesium content in agricultural and food products between Japan and the rest of the world (in Bq/kg)

	Japan (old limit)	Japan (new limit)	Codex Alimentarius Commission	EU	USA
Limits (radioactive cesium)	Drinking water 200 Milk and dairy products 200 Vegetables 500 Grain 500 Meat, egg, fish etc. 500	Drinking water 10 Milk and dairy products 50 Infant food products 50 General foods 100	Foods (except infant food products) 1,000 Infant food products 1,000	Drinking water 1,000 Dairy products 1,000 Infant food products 400 Other food products 1,250	1,200
Concept of limit setting	The radiation dose is 5 Sv or less per year	The radiation dose is 1 Sv or less per year (assume that 50 % of general foods, 100 % of milk and dairy products and infant food products are polluted)	The radiation dose is 1 Sv or less per year (assume that 10 % of food products are polluted)	The radiation dose is 1 Sv or less per year (assume that 10 % of food products are polluted)	The radiation dose is 5 Sv or less per year (assume that 30 % of food products are polluted)

Source: Compiled based on materials from the Agriculture, Forestry and Fisheries Department of Fukushima Prefecture

Unit is Bq/kg

Table 16.2 Attributes of respondents

		2011		2012	
		Fukushima "Aizu farmers' markets" (n=216)	Kanagawa "Ashigara farmers' markets" (n=242)	Reconstruction Aid Event "Fukushima and Tohoku Festival" (n=48)	General Farming Event "Agricultural Frontier" (n=181)
Sex	Male	13.9	28.1	52.1	47.0
	Female	86.1	63.2	47.9	53.0
	No response	–	8.7	–	–
Age (years)	19 or less	–	2.1	–	3.3
	20–29	6.5	5.8	16.7	8.3
	30–39	5.6	12.4	22.9	17.1
	40–49	16.7	17.8	29.2	27.6
	50–59	27.8	18.2	18.8	9.4
	60–69	27.8	27.3	6.3	9.4
	Over 70	13.9	14.0	4.2	7.2
	No response	1.9	2.5	2.1	17.7
Child	With child(ren)	34.7	48.8	27.1	63.0
	No child	65.3	51.2	72.9	37.0

Source: Created from survey by authors

Notes: Figures in the table refer to percentages

Frontier 2012" fair and the other targeting 48 people who attended the Fukushima and Tohoku Festival, an event organized to support the recovery. The two latter events were held in Tokyo from the end of October to December 2012. The surveys addressed how consumer sentiment with regard to radioactivity testing methods had changed since immediately after the disaster. Table 16.2 shows the distribution of the respondents in terms of attributes.

16.2.2 Characteristics of Consumer Sentiment Immediately After the Disaster (2011)

In our 2011 surveys we questioned residents in Kanagawa Prefecture's Ashigara region and Fukushima Prefecture's Aizu region on their opinions with regard to different limits for radionuclide in agricultural products [500 Bq/kg or less, 100 Bq/kg or less, and not detectable (ND) using standard detection devices]. About 40 % to 50 % of the respondents replied that they could not decide if they were reassured by the provisional limits of 500 and 100 Bq/kg that were applied during the first year after the disaster. Because the meaning of the term "provisional limit" had not been explained clearly to consumers in the aftermath of the disaster, they could not judge whether the limits were safe, and doubts about the credibility of the actual figures used for the thresholds were also observed. Although lowering the radioactive

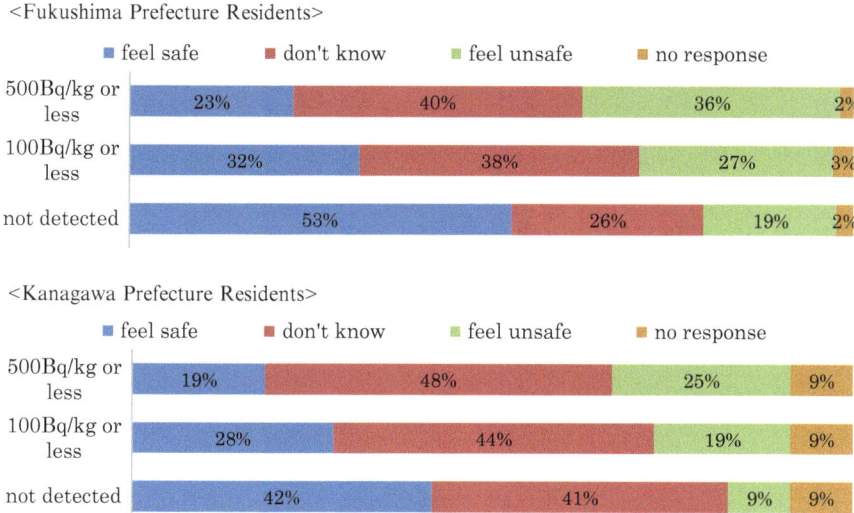

Fig. 16.1 Consumer sentiment toward radioactive content limits for agricultural products (2011) (From survey by authors)

Table 16.3 Consumer sentiment toward radioactive content limits for agricultural products by gender

	Male			Female		
	500 Bq/kg or less (%)	100 Bq/kg or less (%)	Not detected (%)	500 Bq/kg or less (%)	100 Bq/kg or less (%)	Not detected (%)
Feel safe	16	34	51	16	24	40
Don't know	43	40	35	47	46	41
Feel unsafe	35	22	7	25	19	10
No response	6	4	6	12	12	10

Source: Created from survey by authors

content limit for agricultural products from the provisional limit of 500–100 Bq/kg could potentially make consumers feel somewhat safer, nearly one in four consumers replied that they felt anxious about a 100 Bq/kg threshold as well. Furthermore, about 40 % to 50 % replied that they would only feel reassured if the limit were set at the ND level, indicating that a sense of anxiety from lack of appropriate information had aggravated the reputation-based damage. These results indicate that when information lacking in credibility is provided, there is greater ambiguity and significantly greater reputation-based damage (Fig. 16.1).

Considering the results by parental status-based attributes, the cohort of parents with younger children had a higher tendency to feel reassured if radionuclides were not detectable, indicating that most of them were seeking a more stringent criterion. Among the other groups, many replied that they could not decide what they felt about the different limits, indicating feelings of ambiguity over safety (Tables 16.3, 16.4, 16.5).

Table 16.4 Consumer sentiment toward radioactive content limits for agricultural products by age group (%)

	Youth			Middle-aged group			Senior		
	500 Bq/kg or less	100 Bq/kg or less	Not detected	500 Bq/kg or less	100 Bq/kg or less	Not detected	500 Bq/kg or less	100 Bq/kg or less	Not detected
Feel safe	18	35	47	18	30	41	14	22	43
Don't know	41	35	37	44	39	38	49	49	38
Feel unsafe	37	24	12	30	24	14	22	14	5
No response	4	6	4	8	7	7	15	15	14

Source: Created from survey by authors

Table 16.5 Consumer sentiment toward radioactive content limits for agricultural products by parental status

	With Child(ren) (elementary school)			With child(ren) (junior high school)			No child(ren)		
	500 Bq/kg or less (%)	100 Bq/kg or less (%)	Not detected (%)	500 Bq/kg or less (%)	100 Bq/kg or less (%)	Not detected (%)	500 Bq/kg or less (%)	100 Bq/kg or less (%)	Not detected (%)
Feel safe	15	25	52	19	25	36	17	31	43
Don't know	35	29	29	47	44	42	47	44	38
Feel unsafe	40	33	10	22	17	11	25	15	9
No response	10	13	8	11	14	11	11	10	10

Source: Created from survey by authors

16.2.3 Characteristics of Consumer Sentiment One and a Half Years After the Disaster (2012)

According to the 2012 surveys, visitors to the agricultural fair (general consumers) expressed generally less positive sentiment than participants in the event to support recovery (recovery supporters) with regard to all three criteria they were asked to evaluate (a radioactivity limit of 100 Bq/kg or less, testing of every bag of produce, and a radioactivity limit at the ND level). However, more than 70 % of both types of consumers replied that they felt reassured to some degree by all three criteria. Nevertheless, among general consumers, there was a drop-off in feelings of reassurance with regard to the 100 Bq/kg limit, with 9 % replying that it made them feel slightly anxious and another 1 % replying that it made them feel extremely anxious. Overall, however, compared to the 2011 surveys, the 2012 surveys showed a heightened sense of reassurance with regard to the 100 Bq/kg and ND limits. It is evident, therefore, that nearly 2 years after the Great East Japan Earthquake and ensuing nuclear disaster, more consumers were aware of the facts regarding the safety of different levels of radionuclides, and provision of information regarding radioactive contamination was reducing their feelings of ambiguity. Consumers reported feeling most reassured by a radioactive content limit of ND, followed by testing of every bag, and then by a limit of 100 Bq/kg or less, the criterion for shipment currently set by the government (Fig. 16.2).

When residents of the Ashigara region in Kanagawa Prefecture were surveyed on how to assuage anxiety with regard to radionuclides, the most popular response, cited by 32 % of respondents, was that factually accurate media reporting would

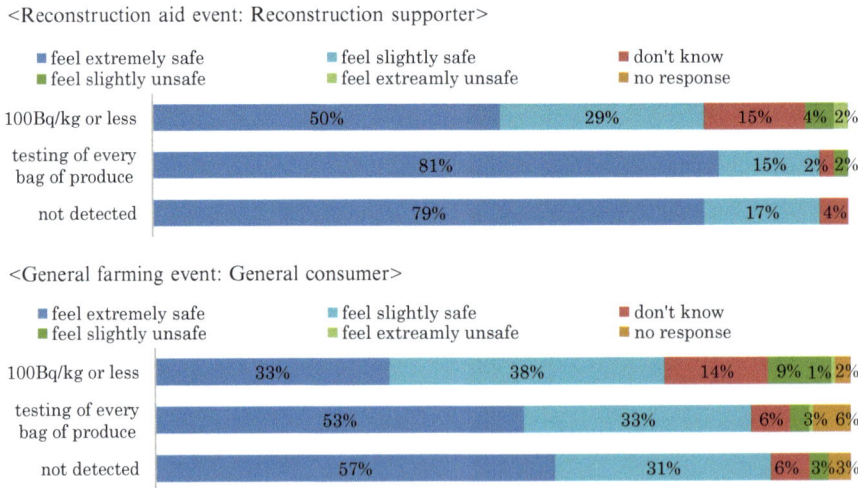

Fig. 16.2 Consumer sentiment toward radioactive content limits and testing method for agricultural products (2012) (From survey by authors)

Table 16.6 Survey results on ways to assuage anxiety regarding radionuclides

Ways to assuage anxiety regarding radioactive substances	Percentage
Factually accurate media reporting	31.9
Easy for consumers to obtain safety-related information	22.2
Testing of every bag of produce	21.5
Maintaining transparency with regard to distribution and the identities of producers	12.5
Others	4.2
No response	7.6

Source: Created from survey by authors

reduce anxiety, followed by 22 % who cited testing of every bag of produce, and another 22 % who replied that making it easy for consumers to obtain safety-related information would help mitigate the anxiety. In addition, 12 % replied that maintaining transparency with regard to distribution and the identities of producers would be useful (Table 16.6). These results indicate that accurate reporting by the media helps assuage anxiety and enhances consumers' feelings of reassurance regarding radionuclides; in other words, provision of appropriate information is indispensable in overcoming reputation-based damage.

16.3   Effectiveness of PR Activities and Events to Support Fukushima's Recovery

In this section we evaluate the benefits of PR activities and events to support recovery in Fukushima, both of which serve as a means of providing appropriate information. Figures 16.3, 16.4, and 16.5 show the results of the 2012 surveys of consumer attitudes toward purchasing agricultural products from Fukushima Prefecture, targeting visitors to the agricultural fair and the event to support recovery. Consumers were given a choice of five potential responses: "I don't mind where products come from," "I buy Fukushima products to support the recovery effort," "I prefer to buy products from other areas if the price is the same," "I buy Fukushima products if they are cheap," and "I never buy Fukushima products." The results indicate that 36 % of the consumers surveyed did not avoid buying products just because they came from Fukushima. In addition, 25 % would buy Fukushima products to support the recovery effort, suggesting that the tendency to avoid agricultural products and other foods from Fukushima had abated since immediately after the disaster. However, the percentages of consumers who never bought Fukushima products, or would buy them only if they were cheap, were 1 % and 24 %, respectively. If we include those who would prefer to buy products from other areas if the price was the same, it is apparent that close to 40 % of the consumers surveyed viewed agricultural products from Fukushima in a different light from products from other areas, indicating that a negative reputation still remained. Another notable finding of the

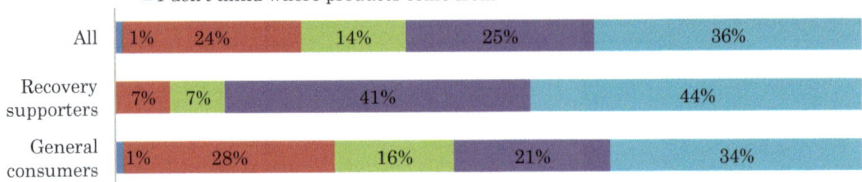

Fig. 16.3 Consumer attitudes toward buying agricultural products from Fukushima (From survey by authors)

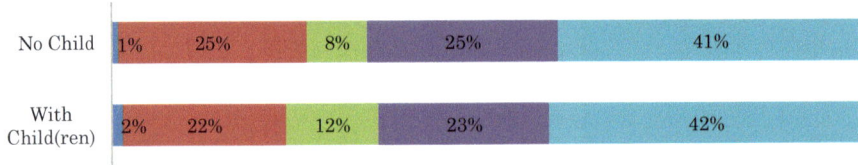

Fig. 16.4 Consumer attitudes toward buying agricultural products from Fukushima by parental status (From survey by authors)

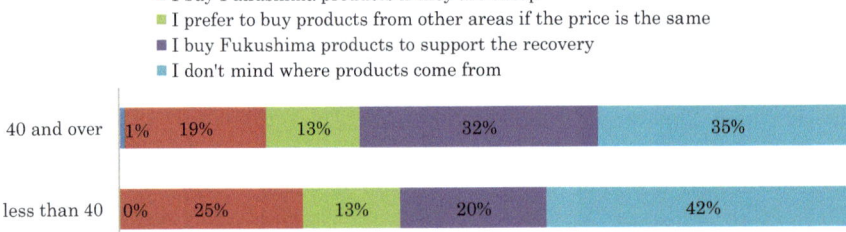

Fig. 16.5 Consumer attitudes toward buying agricultural products from Fukushima by age group (From survey by authors)

surveys was the large number of recovery supporters who replied that they did not mind where products were from, or that they would buy Fukushima products to support the recovery, indicating that securing supporters through PR activities and events to support recovery is effective in overcoming reputation-based damage.

The foregoing results indicate that support for recovery and PR activities may be effective in overcoming reputation-based damage.

16.4   Ways to Overcome Reputation-Based Damage

In this chapter we reviewed previous research on reputation-based damage and initiatives to combat such damage after the disaster. In addition, we elucidated the changes in consumers' feelings of safety and reassurance and the effectiveness of safety-related PR and events to support the recovery. The new knowledge presented in this chapter can be summarized as follows.

First, 2 years have passed since the Great East Japan Earthquake and the ensuing nuclear disaster, and awareness of the facts regarding the safety of different levels of radionuclides has permeated among consumers. At the same time, provision of relevant information is helping to reduce reputation-based damage resulting from consumers' feelings of ambiguity.

Second, securing a base of consumers who buy products to support the recovery has been another effective method of overcoming reputation-based damage. It will be important to continuously implement events to support the recovery going forward.

Third, consumers with children were unable to dispel their anxiety with regard to radioactivity at the 100 Bq/kg level, suggesting the need for more stringent criteria such as setting the level at ND, and testing every bag of produce.

These results indicate that there are effective ways to overcome reputation-based damage relating to agricultural products produced in irradiated areas: (a) provide accurate information on radioactive contamination (dispel ambiguity); (b) use methods that come as close as possible to testing every bag of produce, and require that agricultural products shipped to the market contain no radionuclide detectable using standard detectors (reduce importance); and (c) secure supporters by holding events to support the recovery efforts in disaster zones. It is, moreover, important to implement such initiatives on a continuous basis.

References

Satou T (2004) Rumor and panic. Ritsumeikan Human Sci Res 7:193–203 (in Japanese)
Sekiya N (2003) The social psychology of reputation-based damage: The actual and mechanism of reputation-based damage. J Disaster Inf Stud Jpn 1:78–79 (in Japanese)

Part IV
Activities and Impressions of Students and Farmers Who Supported the Project

Chapter 17
Staking Recovery Hopes on Soma Revival Rice

Kaisei Inagaki, Tomoko Ninagi, Saburo Sasaki, and Akiko Sato

Abstract This chapter conveys the sense of crisis and mission felt by students who participated in soil survey analyses to help restore paddy fields immensely damaged by the tsunami. Also, it presents the background and hard work that led to the harvesting and sale of "Soma Revival Rice". This rice was grown on tsunami-damaged paddy fields recovered through a method developed by the Tokyo University of Agriculture.

Keywords Soma Revival Rice • Soil analyses • Tsunami-damaged paddy fields

17.1   The Story Behind Soma Revival Rice

Kaisei Inagaki
Department of Applied Biology and Chemistry,
Tokyo University of Agriculture, 1-1-1 Sakuragaoka,
Setagaya-ku, Tokyo 156-8502, Japan
e-mail: k3inagak@nodai.ac.jp

It was September 5, 2011, and we were on our way to Fukushima to study the effect of earthquake and tsunami-related damage on strawberry greenhouses. Partway through our journey, Professor Itsuo Goto received an e-mail. It was from Professor Toshiyuki Monma, explaining that farmers in the Iwanoko district of Soma city had requested help in restoring their damaged rice paddies. We changed our plans for the following day and headed for Iwanoko. This was the beginning of the Soma Revival Rice project.

We arrived on the afternoon of September 6 to find four people waiting for us, including Norio Sato, the head of the Iwanoko regional agricultural recovery asso-

K. Inagaki (✉) • T. Ninagi • S. Sasaki • A. Sato
Department of Applied Biology and Chemistry, Tokyo University of Agriculture,
1-1-1 Sakuragaoka, Setagaya-ku, Tokyo 156-8502, Japan
e-mail: k3inagak@nodai.ac.jp

© The Author(s) 2015   235
T. Monma et al. (eds.), *Agricultural and Forestry Reconstruction*
After the Great East Japan Earthquake, DOI 10.1007/978-4-431-55558-2_17

Fig. 17.1 Conducting soil study in a damaged rice paddy

ciation. Soma's Iwanoko district, which looks out onto Matsukawaura Prefectural Natural Park, has 150 ha of rice paddies located 2 km from the coast. Most paddies were littered with tsunami debris, ranging from pine trees uprooted from the sandbanks of Matsukawaura Lagoon, to debris from ruined homes, to boats. Furthermore, the surface of the paddies was covered with sediment that had been dredged from the ocean floor by the tsunami. The lack of a foreseeable way forward had sapped farmers of their will to cultivate the land.

We began by investigating four sites in the Iwanoko district with different levels of damage (Fig. 17.1). We divided the soil into three layers—the tsunami sediment covering the paddy surface, the prepared topsoil originally intended for rice planting, and the subsoil underneath—and analyzed their content. We found that the tsunami sediment washed from the seabed was rich in magnesium and potassium, substances required for plant growth. And although we feared that the sediment may contain heavy metals, our analysis found their level to be no different from that of the topsoil, which meant it had no impact on rice cultivation. On the other hand, the high concentration of salt from the seawater made it impossible to plant crops in the soil as it was.

To address this problem, near the end of September we combined the tsunami sediment and topsoil of a paddy of 60 ares owned by Sato, then created mole drains and attempted to remove the salt from the paddy using rainwater only. Compared to a neighboring paddy field whose layers were not mixed, the salt-removing effect was marked. This observation meant that the tsunami sediment did not need to be removed, and that farmers could restore their paddies on their own using agricul-

Fig. 17.2 Holding a briefing meeting on measures to restore rice paddies

tural machinery. This discovery gave Sato and the other farmers hope that they might be able to recommence farming.

On the evening of November 3, the university gathered the members of the Iwanoko recovery association and held a briefing on recommencing rice farming. It would be up to the local farmers to decide whether to remove the salt from their paddies using the successful Tokyo University of Agriculture Method (also known as the Soma Method). Around 40 participants were in attendance, and they listened intently as Professor Goto used slides to explain the situation. The salt removal method, which differed from the procedure recommended by the Ministry of Agriculture, Forestry and Fisheries, surprised the farmers. After the meeting, which lasted a little less than 2 h, the association members stayed late into the night discussing the best course of action (Fig. 17.2).

In the end, Sato concluded "We won't know unless we try," and at the end of November performed the salt removal process on two more paddies in addition to the original paddy whose soil layers were mixed in September. Nonetheless, only 1.7 ha of rice could be planted in 2012, which brought home to us the difficulty involved in restoring the rice paddies.

The tsunami sediment with which the topsoil was mixed contained pyrite. On the ocean floor, pyrite is stable and does not react. However, when exposed to air it oxidizes to form highly acidic sulfuric acid, lowering the soil pH (H_2O) to around 3. Countermeasures were therefore required to proceed with cultivation. To address this problem, we focused our attention on converter slag, which is available as a

Fig. 17.3 Using a lime sower to spread converter slag

commercial ameliorating agent. In addition to being rich in calcium, converter slag is also rich in iron and silicic acid. Adding this agent to the paddies not only lowers the acidity of the soil, but also prevents aging and hydrogen sulfide toxicity, effectively killing three birds with one stone and making it the logical choice.

The soil in the three paddies spanning 1.7 ha had also acidified. With support from Nippon Steel & Sumitomo Metal Corporation and Minex Corporation, who provided us with converter slag in three different pellet sizes, on April 24, 2012, we used a lime sower to distribute 0.5–1.0 t/10 ares of the agent (Fig. 17.3). As a result, the soil pH (H_2O) rose to a level suitable for rice cultivation.

Although some irrigation and drainage channels were damaged, on May 11 we filled the paddies with water and planted rice seedlings (Hitomebore variety) (Fig. 17.4). A year and 2 months after the earthquake, in a district devastated by the ensuing tsunami, we had managed to conduct our own salt removal process and plant rice paddies amidst surrounding fields that had been considered untouchable and left to run wild. Our paddy of rice seedlings, which lay among overgrown fields as far as the eye could see, became a local center of attention and a symbol of the first step toward restarting agriculture in Soma. With this heavy pressure, failure was not an option.

The city of Soma is located in Fukushima Prefecture, approximately 40 km from the Fukushima Daiichi Nuclear Power Station. Unless the radioactive cesium concentration in the rice harvested in 2012 was within the safety limit (100 Bq/kg) for

Fig. 17.4 Rice seedlings were successfully planted

unpolished rice, we would not be able to sell it. To this end, in August we collected stems and leaves to measure the concentration of radioactive cesium. Our results found the level to be below the limit of detection of 10 Bq/kg. The radiation concentration in unpolished rice grains is known to be around one third to half the level in the stems and leaves. This finding relieved Sato.

On September 27, the first crop of rice was harvested under a cloudless sky. The rice plants gathered by the combine were heavy, and from the outset Sato had a feeling that this year's crop was better than most. Once the plants were transported to his house—where the drier was kept—dried, and hulled, the size of the harvest became evident: 630 kg of rice per 10 ares. Before the earthquake, the harvest of Hitomebore had averaged around 600 kg per 10 ares. Thus, as far as the harvest was concerned, we could not have hoped for better and we all celebrated, but there were inevitable worries about the radiation content. Yet neither Fukushima Prefecture's "every bag" testing nor the university's germanium detector found any trace of cesium, and the rice was confirmed safe for consumption. All in all, the 1.7 ha of paddy fields generated about 10 tons of unpolished rice. Once polished, the rice was given the name "Soma Revival Rice" (Fig. 17.5), and in 2013, the area planted grew to about 40 ha. As a result of this success, I am hoping that the creation of Soma Revival Rice will serve as a first step toward recovery.

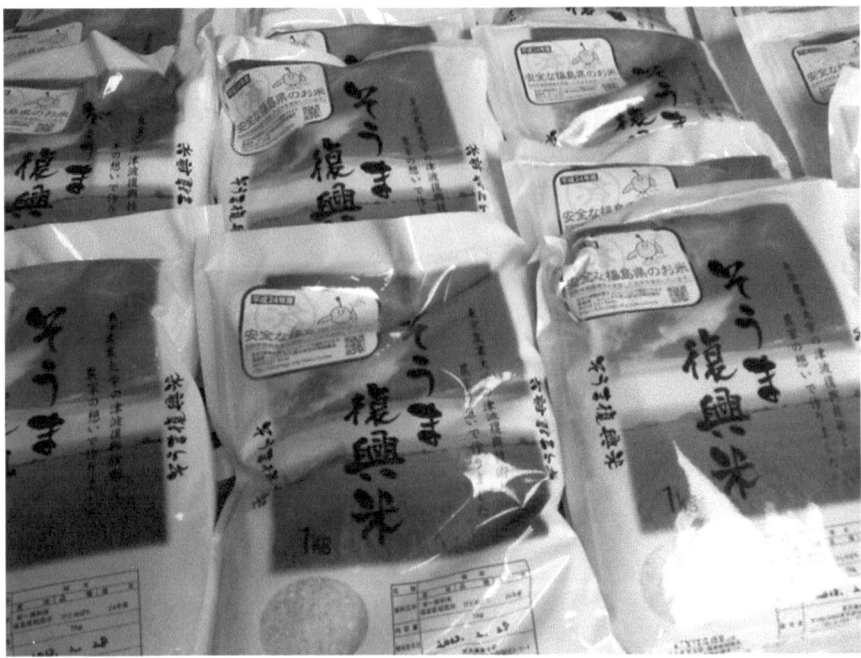

Fig. 17.5 The harvested Soma Revival Rice

17.2 Bonds Form as Rice Grows: Participating in the Revival Rice Project

Tomoko Ninagi
Department of Applied Biology and Chemistry,
Tokyo University of Agriculture, 1-1-1 Sakuragaoka,
Setagaya-ku, Tokyo 156-8502, Japan

At 10 A.M. on November 2, 2012, the first day of the Tokyo University of Agriculture harvest festival, Soma Revival Rice went on sale outside the co-op store on the Setagaya campus. The five to six students in charge of the stall were shouting out, inviting potential customers to stop by. The stall itself was covered with 1 kg bags of rice arranged in lines. It was a simple booth, just a table and a cash register, but the students' handmade signs and posters gave it something of a presence despite its simplicity.

In May 2011, it was decided that the Laboratory of Agricultural Production and Environmental Chemistry, to which I belong, would form a recovery assistance team. Professor Goto explained the situation in the disaster zone, and I was asked if

I would like to assist. I made up my mind to join then and there. Although parental consent was required to participate, my parents were extremely supportive when I raised the issue, urging me to give it everything I had. Shortly after, the eight-member East Japan Assistance Project soil fertilization team was born, composed of researchers and students headed by Professor Goto. After study and analysis of soil conditions, a salt removal method was established, involving mixing the soil and then exposing it to rainwater. Subsequently, in 2012, rice was planted over 1.7 ha of paddy area in the Iwanoko district. We conducted regular studies of the paddies to observe the progress through to harvest. By June, the paddies next to the barren, salt-damaged land on which even weeds had refused to grow were filled with green shoots stretching toward the sky. By August heads of rice had taken shape, and by September a golden carpet had begun to form on the otherwise desolate land. These golden ears of rice, drooping with their own weight, would rustle as the wind swept across the vast plains. The sparkling golden grains shone under the clear fall sky, engraving their image in the minds of all who gazed upon them. Before the heads were harvested by combine, we collected a sample to study. No matter how bountiful the harvest, its safety had to be confirmed. We ran tests to determine levels of radiation, heavy metals, and quality. Our results found no trace of radiation, and heavy metal levels that were far lower than the designated limits. These results proved that the rice was safe and could be consumed with peace of mind, so the decision was made to sell the Revival Rice at the university's harvest festival.

Our sales target was 1 ton of rice. By selling the rice, we hoped to demonstrate that safe, quality-assured rice could be grown in Fukushima. We hoped that when people ate it, they would think not only of how good it tasted, but also of the people who grew it. The walls of the stall were filled with photographs showing the situation in the affected area and explaining the processes that were undertaken to grow the rice. Beside the register were A4-sized flyers explaining the concept behind Revival Rice. Although there were few visitors in the morning and sales of the rice were slow, some people stopped to look at the photos and ask about the product. Then the flow of customers gradually picked up, and our rice began to sell. Customers ranged from young children to senior citizens, with families and elementary school students among those making purchases. There was particular interest from homemakers, who would often question us intently about the safety of the rice. Although we were slightly concerned that the 1-kg size might be too heavy and deter customers, these worries proved unfounded. One woman even returned the next day to buy more, telling us that the rice she had purchased the day before was delicious. There were customers who would tell us, with tears in their eyes, that they were from Fukushima. One person bought more than a dozen bags to distribute to acquaintances. And just like that, our 1,000 bags of rice had sold out (Fig. 17.6).

This crop, harvested from a little corner of the tsunami-damaged region, generated a great deal of interest, and this year 40 ha of paddies have been planted in the Iwanoko district. Norio Sato, the farmer who grew the Revival Rice, told us, with a smile on his face, that it had given him renewed energy. Then he moved straight

Fig. 17.6 Soma Revival Rice is sold at the harvest festival

onto the topic of planting next year's crop. I realized that when people enjoy eating rice that tastes good and that the farmers enjoyed growing, bonds are forged. Each of the stages—from plowing and growing to selling, buying, and eating—is linked, and each of us is responsible for one of those stages. I hope to cherish my role in the process and keep these bonds between farmer and consumer alive.

17.3 Participating in the East Japan Assistance Project

Saburo Sasaki
Department of Applied Biology and Chemistry
Tokyo University of Agriculture, 1-1-1 Sakuragaoka,
Setagaya-ku, Tokyo 156-8502, Japan

I first joined Tokyo University of Agriculture's East Japan Assistance Project during the soil fertilization team's second Fukushima survey in June 2011. Touring the damaged coastal districts of Soma, I was lost for words at the sheer devastation. Houses with only a skeleton of the ground floor remaining, cars reduced to mere lumps of metal, and boats washed inland as far as the residential areas—they all bore witness to the massive power of the tsunami. Farmlands in the coastal districts were covered in sediment brought in by the tsunami, pine trees ripped from the Matsukawaura sandbanks, and masses of wreckage, and mangled tractors and other farm machinery lay scattered. Drainage pump stations had also been destroyed, and as the earthquake had caused the land to subside, seawater had flowed inland, turning the whole area into part of the ocean. I wondered who would even think of attempting to farm again on this ruined land. Nonetheless, having initially been taken aback by the scale of devastation, over the next one and a half years I would find myself deeply moved by the resilience of the local farmers as they successfully managed to start farming again.

My job mainly involved analysis of soil. Every month I would visit Soma to collect soil samples from tsunami-damaged farmland, then take them back to the laboratory to analyze the level of salt and other nutrients. Although collecting and analyzing a large volume of samples each time was in itself quite taxing, the anxiety of knowing that the data I produced would affect recovery work at the site was even more stressful. Realizing that any errors would cause irreparable damage to the project, I performed each analysis with the utmost care, knowing there would be no second chances.

We spent an anxious month after planting the rice seedlings in May 2012. However, by June our worries subsided as we saw that the seedlings had developed strong roots. As time passed in July and August, the fields neighboring our plot began to grow wild. Then herbicides were sprayed to control this runaway growth, and the wilted brown scrub became reminiscent of wintertime, despite it being the middle of summer. Amidst this, however, stood our three paddies, filled with rice shoots growing green and strong. It was a moving sight. And I will never forget the beauty of the way the plants bent in the September wind before the harvest. Although it covered only a small area, the first sight of shimmering gold in 2 years seemed almost majestic. "I want to encourage my fellow farmers by showing that we can still plant and grow rice ourselves, in spite of the tsunami," commented Norio Sato, the farmer who grew the rice. His success proved that our efforts had not been in vain, and gave us renewed hope.

When we sold the rice at the harvest festival, we were initially worried about whether we would be able to sell it all. However, these concerns quickly vanished. Admittedly, quite a few people were worried about radiation, and we did sense the damage that a negative reputation and misinformation can cause, but most visitors were reassured when we explained that the rice had been thoroughly tested. Many offered kind words of encouragement, and expressed their desire to support the affected regions, asking whether the money they paid would go to the grower. The words of appreciation from former Fukushima residents left a particular impression on me (Fig. 17.7).

It is, however, still the case that, although farmland in all areas damaged by the tsunami will likely be restored over the next several years, recovery from the nuclear accident will take much longer. I believe the university should continue doing everything in its power to provide even greater support until all areas

Fig. 17.7 Soma Revival Rice sold out at the harvest festival

damaged as a result of the Great East Japan Earthquake and its associated disasters have fully recovered.

Being involved with the project brought home to me the huge significance of the calling invoked by the word "Agriculture" in the university's name.

17.4   Revival Rice and *Kizuna* (Bonds of Friendship)

Akiko Sato
Department of Applied Biology and Chemistry,
Tokyo University of Agriculture, 1-1-1 Sakuragaoka,
Setagaya-ku, Tokyo 156-8502, Japan

Initially, after the earthquake, the tsunami footage replayed over and over on television seemed like a scene from a far-off place. However, a year later I joined the earthquake recovery voluntary activities sponsored by the university's parents' association, wanting to see the situation in the affected region with my own eyes. In Ishinomaki, Miyagi Prefecture, the sight of debris still piled high and houses

Fig. 17.8 The plants were still green in mid-August, with ears of rice beginning to form

reduced to their foundations left me at a loss for words. Later, we volunteered at temporary housing units in Higashi-Matsushima, also in Miyagi Prefecture. Although our time there passed by in a flash, the parting words of one of the residents left a particularly lasting impression: "You said that you haven't been able to help us much, but that's not true. Knowing that you haven't forgotten us is all the support we need."

Through this volunteering experience I learned about Tokyo University of Agriculture's East Japan Assistance Project. I was curious to know how my research group was involved, and when I expressed my desire to accompany the team on their survey of Fukushima, I was allowed to participate despite being only a third-year student.

The study took place in early August, and the rice plants were still green, with the ears only just starting to form (Fig. 17.8). I tagged along with the more senior students, helping perform growth studies and collect soil samples. At first I felt satisfied simply being in Fukushima. But as the study progressed, I heard the passion in the farmers' voices as they addressed Professor Goto and research fellow Kaisei Inagaki, and realized that to assist the recovery effort I first needed to understand the problems people were facing. I needed to study harder.

I also helped sell Soma Revival Rice at the harvest festival. I believe that selling the rice helped deliver hope to the farmers in the disaster zones as well as increase the number of people supporting them as they worked to recover. I intend to continue supporting the afflicted region by remembering the people impacted by the disasters, visiting Tohoku and buying Tohoku-made products, and studying and researching the situation with them in mind.

Chapter 18
Impressions of the Students Who Participated in the Radioactivity Monitoring System of Farmland

Volodymyr Ganzha, Keiji Kanamori, Hana Fujimoto, and Ryo Itakura

Abstract This chapter conveys the sense of crisis and mission felt by students who participated in the development of a radioactivity monitoring system that now covers each farm field to help the farmland to recover so that farming could resume. Also, it depicts the students' hard work in severe conditions and their interactions and shared feelings with the local people.

Keywords Radioactive contamination • Radioactivity monitoring system • Resuming farming

18.1 Meeting Cheerful Farmers and Finding a Second Home

Volodymyr Ganzha
Embassy of Japan in Ukraine,
4, Muzeiny Lane, Kyiv 01901, Ukraine
e-mail: volodymyr2009@hotmail.com

I am conducting research on recovery measures implemented for Ukraine's agriculture and forestry industries in response to radioactive contamination, and I hope to obtain results that will also aid recovery in Fukushima. Doing so requires not only that I read books and other written materials, but also that I travel to the disaster sites repeatedly to talk to the farmers, analyzing the information I gather from a variety of sources to develop new theories and study the validity of existing theories.

V. Ganzha (✉)
Embassy of Japan in Ukraine, 4, Muzeiny Lane, Kyiv 01901, Ukraine
e-mail: volodymyr2009@hotmail.com

K. Kanamori • H. Fujimoto • R. Itakura
Department of International Biobusiness Studies, Tokyo University of Agriculture,
1-1-1 Sakuragaoka, Setagaya-ku, Tokyo 156-8502, Japan

© The Author(s) 2015   247
T. Monma et al. (eds.), *Agricultural and Forestry Reconstruction*
After the Great East Japan Earthquake, DOI 10.1007/978-4-431-55558-2_18

Measuring the surface dose of soil and space dose

Although I do not remember exactly how many times I have visited Soma's Tamano district since April 2012, I think it has been at least ten times. During 2012, in particular, I often participated in studies to develop a monitoring system for radioactive materials in the farmland of Tamano. Measuring radiation levels in the pastures there involved tramping around a large area of sloping land carrying a lead shielding cylinder for blocking background ambient radiation that weighed around 20 kg. It was exhausting work, but despite the difficulties, we also had plenty of fun moments. The farmers would give us tempting homemade snacks, and on hot days they would often provide us with cold drinks. I really enjoyed that part of it.

Our group had no knowledge of the local geography, so at the outset of the monitoring study the officials who headed each of the four areas within Tamano kindly guided us around each area. They told us who owned and farmed each plot of land, its size, and the history of the crops grown on it. I was surprised at how much they knew about the land use and agriculture in their areas; they were like walking encyclopedias.

During the study we cooked our own breakfast, lunch, and dinner. Preparing and eating our meals together helped us develop a stong sense of camaraderie, and I felt that I understood what Japanese people mean when they talk about relationships forged by "sharing rice from the same pot." As a non-Japanese who had lived mainly in Tokyo, it was difficult for me to understand everything the farmers in Tamano were saying in their local dialect, but their warm smiles, mannerisms, and gentle way of speaking reminded me of farmers in my homeland of Ukraine.

I think that if there is anything a Ukrainian such as myself can do to make a meaningful contribution to recovery from the Great East Japan Earthquake, it is to communicate as accurately as possible the initiatives taken in Ukraine and Belarus in response to the Chernobyl nuclear accident, and the problems still faced there. This is the aim I have in mind as I work on my research.

18.2   Realization That Recovery Is a Long, Repetitive Process

Keiji Kanamori
Department of International Biobusiness Studies,
Tokyo University of Agriculture, 1-1-1 Sakuragaoka,
Setagaya-ku, Tokyo 156-8502, Japan

After the Great East Japan Earthquake I wanted to do something—felt that I *had* to do something—for the Tohoku region where I grew up. But what practical steps should I take to help Tohoku? What *could* I do? I wrestled with these thoughts, unable to take action. Then, when I entered my third year and joined Professor Toshiyuki Monma's research group, he told me that the university was conducting a disaster recovery project in the city of Soma in Fukushima Prefecture. "This is it!" I thought, and made up my mind to join in.

Transportation of shielding tool of lead in excess of 20 kg is hard work

The site I visited was the Tamano district of Soma, located next to Iitate, a village designated as an evacuation zone. My first assignment was measuring radiation levels in Tamano's cultivated land, field by field. We took turns lugging around a 20-kg lead shielding cylinder and measuring the radiation dose inside it. In each field, we measured radiation levels at 1 m and 10 cm above the soil, also taking other measurements such as the depth of the topsoil. It was hard work hauling the 20-kg shielding cylinder around the steep grassland slopes and sprawling fields as we went. The slow, laborious process continued until evening, and by the time we had finished my entire body was exhausted. I remember how good Professor Monma's Thai curry tasted that night.

After we finished measuring the ambient radiation dose in the fields, our next task was collecting soil samples from each field and measuring the radioactive substances they contained. The process of drying the collected samples, breaking the clumps into fine dirt, and running them through the measuring machine was very long and arduous. During the process we had to sit in one place and just pound sample after sample into small pieces using hammers. Dust from the crushed soil would get in our mouths, and I found it really tough as I am not good at sitting still.

I had imagined that radiation testing was a difficult process requiring highly sophisticated machinery. In actual fact, however, it proved to be a slow, repetitive process that required us to physically walk around each field measuring the radiation dose, as well as to collect samples and prepare them for testing. I realized that the best way to approach the recovery effort is to just build up gradually by repeating small tasks such as these.

18.3   Farmers' Hard Work Ruined: The Pain of Reputation-Based Damage

Hana Fujimoto
Department of International Biobusiness Studies,
Tokyo University of Agriculture, 1-1-1 Sakuragaoka,
Setagaya-ku, Tokyo 156-8502, Japan

This was the second time I had visited Soma in Fukushima Prefecture as a volunteer. The first was in September 2011, 6 months after the Great East Japan Earthquake. The local *nashi* (Asian pear) farmers faced a labor shortage because of the earthquake and associated disasters, and I helped them to thin their crop. At that time, the farmers told me that what worried them most was whether the pears they had worked to grow would sell in the wake of the Fukushima Daiichi nuclear accident. Then, in June 2012, more than a year after the disasters, the university began measuring radiation levels in the Tamano district to develop decontamination measures. I volunteered to participate.

Checking the field position is one of the important tasks

Soma's Tamano district is located next to the village of Iitate, which was completely evacuated after high levels of radiation were found. The government had prohibited any planting of rice crops for consumption in the area during 2012. So, to be honest, I was anxious until I actually traveled to the site. What was it like in a high-radiation area? I imagined that the residents wore face masks all the time, that people were scared of venturing outdoors, and that the whole region was desolate and empty. But I was relieved to find that everyday life continued more normally than I had expected. At the same time, however, I noticed that the pastures had been left to run wild and were overrun with growth. Seeing this made me realize how severe the radioactive contamination had been and how much the local farmers had suffered.

In my opinion, the biggest current hurdle to recovery from the disasters is the problem of economic damage from a negative reputation. It was when we sold the Soma Revival Rice that I really felt this firsthand. Soma Revival Rice is grown in severely tsunami-damaged paddies in Soma's Iwanoko district after cleansing the soil using a salt removal method christened the Tokyo University of Agriculture Method. The rice had been tested with a germanium semiconductor detector and confirmed safe for consumption. As we offered it for sale, many people were positive, saying that they wanted to help Fukushima to recover and were willing to eat the rice if it was safe. Others, however, said that even if it had tested safe, they did not want to buy it because it was from Fukushima and they still felt uneasy. As this experience shows, the unease toward agricultural products from Fukushima and the associated reputation-based damage remains, even more than 2 years after the accident.

The farmers of Fukushima are investing painstaking care in decontaminating the area and ensuring the crops meet radiation limits before finally sending them to market. They are passionate about their profession, saying they do not care if their crops sell for far less than they used to, just so long as people eat them. This is why it saddens me to see people look at the crops these farmers have put their hearts into growing and say "I'm not buying it if it's from Fukushima," or "Even if it meets the official safety limit, it's still not safe." Rather than simply saying that Fukushima is dangerous, I wish people would ask themselves what we should do about it. I wish people would stop destroying the hopes of producers, who have worked even harder than usual from a desire to see people eat and enjoy produce that meets the required safety limits.

Tamano is a fantastic place developed over the years by wonderful people. When we were there, measuring the radiation in each field, we needed information such as details of the sites, the plot sizes, and the owners' names. The officials and farmers of each area in Tamano accompanied us in the searing heat as we worked, patiently explaining what we needed to know at each site. They would always provide us with plenty of drinks, telling us not to get heatstroke, and when we visited their homes to eat lunch, they would welcome our dirt-encrusted team with a smile. That is why I have decided to provide what assistance I can so that the residents of Tamano may plow their paddies and fields with peace of mind, and resume living the life to which they are accustomed.

18.4 "I Want to Farm Again": The Farmer's Words I Can't Forget

Ryo Itakura
Department of International Biobusiness Studies,
Tokyo University of Agriculture, 1-1-1 Sakuragaoka,
Setagaya-ku, Tokyo 156-8502, Japan
I have participated in the Soma Project on three occasions to date. These experiences have been more valuable than I ever could have imagined and have had a major influence on my life at the university.

Students rest in the shade of a tree

The first two times I helped with the assistance activities, it was before the radioactive contamination had been removed, which meant working in paddies, fields, and pastures with a slightly elevated radiation dose. It was physically exhausting work as we had to carry a heavy lead shielding cylinder to accurately measure radiation levels by excluding background radiation. Our study also coincided with the rainy season, so the soil in the fields turned to slush, and we would occasionally lose our footing. It was so tiring that on the first day when we arrived back at our accommodation base I climbed straight into my futon and slept.

When I was measuring radiation levels, the numbers on the dosimeter I carried with me were a constant reminder of the radiation fear. It felt as though radiation—the so-called "invisible fear"—was there right in front of my eyes. The abandoned, overgrown pastures and the unplowed fields seemed out of place even to me, a stranger viewing them for the first time. And people were living in this environment. Seeing that made me realize for the first time that we really were in a disaster area. In between times of work, I also had the opportunity to talk to Tamano's farmers, who helped us with our study. Speaking directly with the people involved gave a far more vivid picture of what was going on in the disaster zone than reports on television and other media. Just hearing real stories of the disasters moved me to tears. And the sight of the farmers battling to continue farming under such harsh conditions made a deep impression on me. I cannot forget the words of one farmer who told me that he hoped they would be able to starting farming again the following year.

The March 2011 disasters caused a range of unprecedented problems, including economic damage from negative reputation. As a consequence, restoration of the farmland alone will not be sufficient for agriculture to recover. In addition to rehabilitating agricultural land, we must consider what can be done to conquer reputational damage and allow people to once again buy produce from the affected regions with peace of mind. Participation in the Soma Project served as the first step for students such as I to encounter these problems firsthand, form our own opinions, and consider what we can do to help.

Although Tokyo University of Agriculture runs other programs to assist the affected regions, the Soma Project is focused especially on agricultural recovery. Among various initiatives, the Soma Project is particularly characteristic of the university, and is a program in which students can feel that their efforts are helping revitalize the affected region. There are not many opportunities for students to experience firsthand the agriculture, conditions, and lives of people in the disaster zone, and I believe it is the perfect opportunity for those wanting to make a contribution to the region's revival. I hope that the recovery keeps moving forward, step by step, and I want to keep doing whatever I can to help as a volunteer—no matter how small my role may be.

Chapter 19
Impressions of the Forestry Managers and Students Who Participated in the Radioactivity Damage Investigation of Forests

Eihachi Horiuchi, Kiyoaki Sasaki, Masaaki Itakura, Chisato Yasukawa, and Chihiro Kinoshita

Abstract This chapter portrays the distress of owners and workers of radiation-contaminated forests and the sense of fear and mission felt by students who participated in forest field surveys in such severe environments.

Keywords Radioactive contamination in forests • Forest owners • Forestry cooperatives • Measuring radioactive content of trees

19.1 The Ordeal Inflicted by the Great East Japan Earthquake and the Spread of Radiation

Eihachi Horiuchi
Soma Regional Forestry Cooperative, Nishiki-cho, Haramati-ku,
Minamisoma, Fukushima 975-0031, Japan

Soma's regional forestry cooperative is located in the north of Fukushima's Hamadori region and is responsible for an area spanning two cities and one town. Our cooperative, which has a membership of around 2,600, works to

E. Horiuchi (✉)
Soma Regional Forestry Cooperative, Nishiki-cho, Haramati-ku,
Minamisoma, Fukushima 975-0031, Japan

K. Sasaki
Temporary house, Ushigoe, Haramati-ku, Minamisoma, Fukushima 975-0017, Japan

M. Itakura • C. Yasukawa • C. Kinoshita
Department of Bioscience, Tokyo University of Agriculture,
1-1-1 Sakuragaoka, Setagaya-ku, Tokyo 156-8502, Japan

© The Author(s) 2015   255
T. Monma et al. (eds.), *Agricultural and Forestry Reconstruction
After the Great East Japan Earthquake*, DOI 10.1007/978-4-431-55558-2_19

conserve forests and increase production output in the approximately 34,000 ha of forest (approximately 22,000 ha of which is privately owned) in the area. The Great East Japan Earthquake of March 11, 2011 led to explosions at the Tokyo Electric Power Company (TEPCO) nuclear power station, which is located to our south. Because of the myth of safety surrounding nuclear power, nobody in their worst nightmares imagined that radiation would spread or that people would be forced to evacuate, a mindset that led to secondary damage. Two to three days after the accident, all Minamisoma's residents were ordered to evacuate, including those living outside the 30-km zone. All facilities serving the public were closed, with the exception of government organizations such as municipal offices, and major panic ensued as distribution of products came to a halt. When the spread of radioactive material began to ease from around March 20, we began recovery work such as searching for missing persons outside the 30-km zone and using forestry machinery to clear debris (we also requested assistance from other cooperatives.)

At the beginning of July, when the situation was looking dire, Professor Takahisa Hayashi of Tokyo University of Agriculture visited our office and offered to conduct a study of radioactive contamination from the nuclear accident in forests and trees. We gratefully accepted. As a result we learned that the soil, litter layer, and trees were contaminated with high concentrations of radioactive cesium, and we appealed to Fukushima's prefectural forestry association and other cooperatives, pointing out that the state of radiation contamination in forests should be studied and ascertained. We held briefings on damage compensation at each cooperative under the prefectural association, and are currently in the process of filing for compensation. We also appealed to other organizations such as the Forestry Agency and the Forestry and Forest Products Research Institute, pointing out that the presence of contamination hot spots in mountain forests meant that ongoing and detailed studies were needed. In March of the following year the Forestry Agency and the Forestry and Forest Products Research Institute finally conducted studies at 12 sites in Fukushima Prefecture, and in August the Forestry Agency released the study results—just once. The highest concentration of cesium was 497 Bq/kg, measured in Japanese red pine sapwood in the Ohara district of Haramachi in Minamisoma. As a high ambient radiation dose in the forest corresponds to a high cesium concentration in the bark and wood, we also requested investigations at a further two sites. However, this request was turned down.

About 10 months later, when Professor Hayashi retested Japanese cedar, Japanese red pine, and other trees in forests in the same Ohara district in which the Forestry Agency study was conducted, some trees were found to have around three times the cesium previously reported. Subsequently, Professor Hayashi leased an area of forest in the Ogai district of Haramachi, Minamisoma, where he has planted a variety of saplings to study the absorption of radioactive material during their growth stages. He is also conducting ongoing trials that involve sprinkling water-dispersible potassium chloride and ammonium nitrate powder on trees via radio-controlled helicopter and monitoring their radiation absorption.

However, there has been no progress on compensation for the damage to our forests because TEPCO has not set compensation standards for forest-related damage. Our young successors are despondent. Despite having inherited these assets from their forebears, the radiation-contaminated wood cannot be used, and their forests have lost their value. Looking ahead, I am worried about how to develop plans for rehabilitating, consolidating, and operating our forests as a business, and whether we will be able to obtain the understanding and cooperation of forest owners.

19.2   An Oasis in the Desert

Kiyoaki Sasaki
Temporary house, Ushigoe, Haramati-ku,
Minamisoma, Fukushima 975-0017, Japan

I previously lived in a small village 16 km from the nuclear power station. It has now been designated a long-term evacuation zone, and the region around the village has been designated a potentially permanent evacuation zone. I own a forest of approximately 40 ha, which my family has worked hard to raise and beautify over generations, in between farming. We planted pine trees on the flat land, and although that pine forest was our pride and joy, it suffered damage from pests. After a range of measures, including changing tree varieties, our trees finally reached the pruning stage. The woods surrounding our house cover approximately 1 ha, and are now made up of giant trees of the Japanese cedar, Japanese zelkova, and *momi* fir varieties.

However, this fruit of my ancestors' labor was instantly made worthless by the explosions at the nuclear plant. Although crop and livestock farmers suffered similar damage, they have strong backing from agricultural and livestock industry groups, local governments, and government bodies, and I hear that they have managed to win compensation. However, forestry was the only sector in which industry groups and related organizations stayed silent, and we could not even get a proper investigation into the extent of contamination. The days I spent in Fukushima City after being evacuated were filled with despair.

In January 2012, nearly a year after the nuclear accident, a legal advisor for the Fukushima prefectural forestry association held a briefing session on compensation issues. At the same meeting, Professor Takahisa Hayashi from Tokyo University of Agriculture explained the results of studies into the extent of contamination. I did not know anything about the science involved, so I asked him straight out whether our trees could no longer be sold as products. Professor Hayashi's reply was "That's probably the way it's going to end up." I got the impression that the legal advisor's explanation was not in line with the actual damage that had occurred.

Naturally, there was no proactive discussion from the association, and it seemed as though they were deliberately being vague about the real state of the damage. I traveled to the association's headquarters in Fukushima City together with Takeyama, another forestry operator, and we asked them to reconsider their policy. They replied that they had recently made an approach to TEPCO regarding compensation, but there have been no subsequent developments, and I have doubts about how serious they really are.

Given these circumstances, we have been gathering volunteers since last year and have decided to take action. If the worst comes to the worst, we are considering filing an independent lawsuit. But it is impossible for untrained people to measure the contamination levels. Thankfully, Professor Hayashi informed us that it is Tokyo University of Agriculture's policy to proactively study contamination levels free of charge. As a result, we have benefited from highly dedicated investigation of those levels right through to the present. It has been like finding an oasis in the desert. If it develops into a court case and we win compensation, I hope to donate to Professor Hayashi's research group.

19.3 The Difficulty of Assessing Radioactive Contamination in Forests

Masaaki Itakura
Department of Bioscience, Tokyo University of Agriculture,
1-1-1 Sakuragaoka, Setagaya-ku, Tokyo 156-8502, Japan

From our base in Minamisoma, we conducted studies of forests in the town of Shinchi and the cities of Soma and Minamisoma. In contrast to its support for regular farmers, the government's support for local forestry operators was given a low priority, and operators were unable to ascertain the degree of contamination to the trees they had often spent 50 years or more raising, data that are required to receive compensation. We therefore provided local forestry operators with information on this invisible radioactive contamination in the form of accurate values and tried to help them obtain the appropriate compensation. To do this, it was essential that we accurately identify the extent to which the trees were contaminated.

One fact that became clear from studying trees in various districts was that forest contamination was not proportional to the ambient dose. In some areas we found highly concentrated contamination in the wood despite a low air reading, and vice versa. However, at present the ambient dose is used to determine forest contamination, and in some cases forestry operators are not able to receive compensation for their trees despite them being too contaminated to ship to market. We thought, therefore, that to accurately identify the radiation level of all of Fukushima Prefecture's forests, we needed to take dosage measurements of trees and soil in subdivided plots. This realization brought home the difficulty of identifying contamination levels in vast forest areas. As the field of study was radiation, which was not our area of specialization, the field work was initially one ordeal after another.

The first hurdle was the fact that, in contrast to the crop fields, we faced a harsh working environment, which included steep slopes, plants that could cause skin irritation—such as Japanese sumac trees—or thorns, and insects such as hornets and horseflies. One of our students was unable to walk for 3 days after being stung by an insect. Experiencing this work environment firsthand quickly made me realize that decontaminating the area using human labor was not realistic. I felt a strong need to consider biological methods of decontaminating the area.

Another surprise was the strength of the radiation dose. In some forest areas it was more than ten times higher than in urban areas. Working in these areas brought constant concern about radiation exposure. To dispel this worry I remember telling myself that a return flight from Tokyo to New York would expose me to about 200 µSv of radiation but working in the forest for 3 days would expose me to only 30 µSv at most, so it was no great concern. It was a shock when my girlfriend at the time asked me not to go to Fukushima.

But I lied, telling her I was going to the university's research forest before heading to Fukushima. I do not regret doing this. As a matter of fact, when our personal internal radiation exposure was measured using a full-body counter, none of the students, including myself, returned abnormal readings.

Consequently, while I was still a student, I was able to experience a crisis situation such as this up close, take action, and learn the importance of facing up to difficulties. Forests act as a dam for radioactive material and I believe that assessing their contamination will be essential for the restoration effort in Fukushima to go forward.

19.4 Studying Forest Radiation Contamination Changed My Outlook on Life

Chisato Yasukawa
Department of Bioscience, Tokyo University of Agriculture,
1-1-1 Sakuragaoka, Setagaya-ku, Tokyo 156-8502, Japan

In July 2011, as a third-year student, I began researching forest restoration, working in the forests of Minamisoma, Fukushima Prefecture. Although our task was forest restoration, in reality this initially involved confirming the level of contamination in each area. We trudged around the nearby woodland, survey meters in hand. There were large areas of contamination, even in the peaceful, refreshing woodland, and I felt the invisible fear of radiation. I was motivated by a desire to see the situation with my own eyes, and to do something to help. I felt the pain of families split apart and people losing their homes, and the distress of seeing the trees one had lovingly raised over many years contaminated with radiation.

I studied ways to decontaminate affected forests as quickly as possible. Our experiments confirmed that feeding nitrogen to plants in a nutrient-poor medium such as forest soil caused them to absorb a large amount of cesium. After conducting forest experiments with a range of tree varieties, we hypothesized a model in which planting certain tree types would help complete the decontamination process more quickly than simply leaving the forest as it was. Meanwhile, my autoradiography experiments to observe cesium in the vascular bundles of plants ended in failure time after time, and I felt like crying from frustration. Although the professors are demanding when it comes to experiments, I was able to keep going thanks to the support of my peers and the more senior students. I also gave slide-based presentations at the annual meeting of the Japanese Society of Plant Physiologists when I was a third-year student, and at the annual meeting of the Japan Wood Research Society when I was in my fourth year. While I gained much from performing practical-science-based research, therefore, I also realized how academia and science are closely linked to society and the world at large.

A couple of years on from the nuclear accident, people opposed to the country's nuclear reactors being restarted hold protests in Tokyo on an almost weekly basis. I have begun attending these demonstrations. The ideals of the participants connected directly with me, and I have become strongly opposed to nuclear power. To be honest, it took the nuclear accident for me to take an interest. The event awakened me to a new way of life. Although it is possible to live life without being aware of important issues, as members of society I believe we should consider issues from our own perspective, make choices, and be prepared to speak up if necessary.

My research led me to think about Fukushima, about radioactive cesium and our forests, and about society. As a member of the generation responsible for tackling Japan's radiation problems, I hope to put this experience to good use in future.

19.5 Finding a Way to Live with Cesium

Chihiro Kinoshita
Department of Bioscience, Tokyo University of Agriculture,
1-1-1 Sakuragaoka, Setagaya-ku, Tokyo 156-8502, Japan

I was in my first year of university when the Great East Japan Earthquake struck. The tsunami that followed the earthquake was a terrible tragedy in itself, and then the radioactive contamination from the nuclear disaster was even more of a concern. I spent days in anguish about what to do—I considered volunteering, then found myself questioning what I could possibly offer in the way of assistance, and wondering whether I was just seeking my own self-satisfaction. The problem was—and this is perhaps not the best way to put it—my image of volunteer work consisted of groups pulling up in buses just to do one-off, simple work, feel pleased with themselves, and pretend to be chummy with each other.

But after being assigned to the Laboratory of Plant Molecular Genetics, I learned about the initiatives the laboratory was undertaking in Fukushima and plunged straight into the action. What awaited me was a world of practical science far removed from the "pretend" one-off volunteer work I had imagined. Our research aimed to prevent the transfer of radioactive cesium to fruit in the city of Date in Fukushima Prefecture, and communication with the professors and the more experienced students at the work site was as real as it could be. We performed experiments jointly with representatives from JA Datemirai (a local agricultural cooperative) and companies. As we worked in the fields, members of the public passing by would stop and talk to us. At first I could not understand what they were saying because I did not know the Tohoku dialect, but as they continued to stop and offer us encouragement, I gradually got used to the local accent.

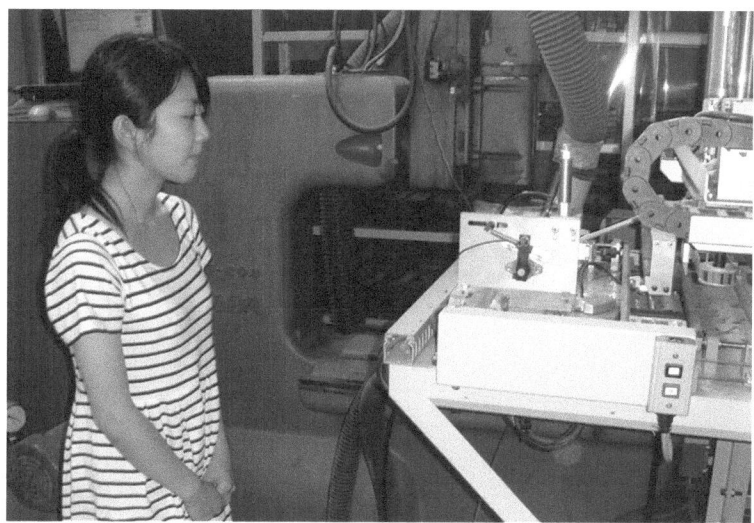

I believe that the research we are performing in the field is really important. We are attempting techniques that, if they prove successful, can be applied on a global level in a range of countries and regions. Our efforts involve several experiments, including spreading pectin on the inner bark of persimmon trees to capture cesium, and spraying leaves or injecting trees with potassium solution to drive out cesium. Whenever we return to the Tokyo campus, we have samples to measure. The other students do not talk when they work, so I let them concentrate by trying to look at the overall picture and do whatever is required to help. I believe that I am getting a chance to learn about radioactive cesium through first-hand experiments and to study how humans can coexist with radiation.